A GIANT WHITE PINE STRADDLING A ROCK IN CONNECTICUT'S CATHEDRAL PINES

EARLY SUMMER MORNING AT ZEALAND POND, IN NEW HAMPSHIRE'S WHITE MOUNTAINS

MAINE'S MOUNT KATAHDIN, SEEN ACROSS A HILLSIDE IN FALL

DEER BROOK RUNNING AMONG GRANITIC BOULDERS, NEW HAMPSHIRE

WHITE BIRCHES IN AUTUMN, BAXTER STATE PARK, MAINE

JUNE-BLOOMING MOUNTAIN AVENS ON NEW HAMPSHIRE'S MOUNT WASHINGTON

SNOW-CLAD EVERGREENS ON VERMONT'S MOUNT MANSFIELD

SKUNK CABBAGE IN A SPRING STREAM, THE BERKSHIRES, MASSACHUSETTS

TIME
LIFE
BOOKS ®

THE TIME-LIFE LIBRARY OF BOATING
HUMAN BEHAVIOR
THE ART OF SEWING
THE OLD WEST
THE EMERGENCE OF MAN
THE AMERICAN WILDERNESS
THE TIME-LIFE ENCYCLOPEDIA OF GARDENING
LIFE LIBRARY OF PHOTOGRAPHY
THIS FABULOUS CENTURY
FOODS OF THE WORLD
TIME-LIFE LIBRARY OF AMERICA
TIME-LIFE LIBRARY OF ART
GREAT AGES OF MAN
LIFE SCIENCE LIBRARY
THE LIFE HISTORY OF THE UNITED STATES
TIME READING PROGRAM
LIFE NATURE LIBRARY
LIFE WORLD LIBRARY
FAMILY LIBRARY:
HOW THINGS WORK IN YOUR HOME
THE TIME-LIFE BOOK OF THE FAMILY CAR
THE TIME-LIFE FAMILY LEGAL GUIDE
THE TIME-LIFE BOOK OF FAMILY FINANCE

NEW ENGLAND WILDS

THE AMERICAN WILDERNESS/TIME-LIFE BOOKS/NEW YORK

BY OGDEN TANNER

AND THE EDITORS OF TIME-LIFE BOOKS

Published simultaneously in Canada. Reprinted 1975.
Library of Congress catalogue card number 73-92887.

THE AMERICAN WILDERNESS
Editorial Staff for *New England Wilds:*
EDITOR: Charles Osborne
Picture Editor: Jane D. Scholl
Designer: Charles Mikolaycak
Staff Writers: Simone D. Gossner, Sam Halper, Susan Hillaby
Chief Researcher: Martha T. Goolrick
Researchers: Paula Arno, Terry Drucker, John Hamlin, Beatrice Hsia, Janice Pikey, Suzanne Wittebort, Editha Yango
Design Assistant: Vincent Lewis

Editorial Production
Production Editor: Douglas B. Graham
Assistant Production Editors: Gennaro C. Esposito, Feliciano Madrid
Quality Director: Robert L. Young
Assistant Quality Director: James J. Cox
Copy Staff: Eleanore W. Karsten (chief), Marilyn Minden, Barbara Quarmby, Florence Keith, Pearl Sverdlin
Picture Department: Dolores A. Littles, Joan Lynch
Traffic: Carmen McLellan

Valuable assistance was given by the following departments and individuals of Time Inc.: Editorial Production, Norman Airey; Library, Benjamin Lightman; Picture Collection, Doris O'Neil; Photographic Laboratory, George Karas; TIME-LIFE News Service, Murray J. Gart; Correspondent Frank Sleeper, Portland, Maine.

The Author: Ogden Tanner, a senior text editor of TIME-LIFE BOOKS, has New England roots in the rocky hillsides of northwestern Connecticut, where his ancestors settled, and for years has skied and hiked the mountain regions farther north. A former newspaper writer and assistant managing editor of *Architectural Forum* magazine, he has been a writer and editor for TIME-LIFE BOOKS volumes on nature, science, history, gardening and photography.

The Cover: Aflame against a dark peak in Maine's Baxter State Park, a sugar maple embodies the brilliance of a New England autumn. Abundant throughout the region, the tree blazes its brightest in a gentle fall of sunny days and cool, snappy nights—the same conditions that stimulate its sap to run sweet and copiously in spring; a specimen this size might yield as much as six pounds of pure maple sugar.

Contents

1/ Stone Walls and White Pines 20
A Winter Calendar 32
2/ The Mountain Roots 44
A Nature Walk to Lost Pond Bog 60
3/ New England's Rooftop 72
Such Lovely, Savage Hills 88
4/ Varmints, Loons and Whiskey Jacks 104
A Fledgling Generation 118
5/ A Private Wilderness 126
The Bangor Tigers 136
6/ The Maine Woods 146
Katahdin's Autumn 166

Bibliography 180
Acknowledgments and Credits 181
Index 182

A Settled Country Growing Wilder

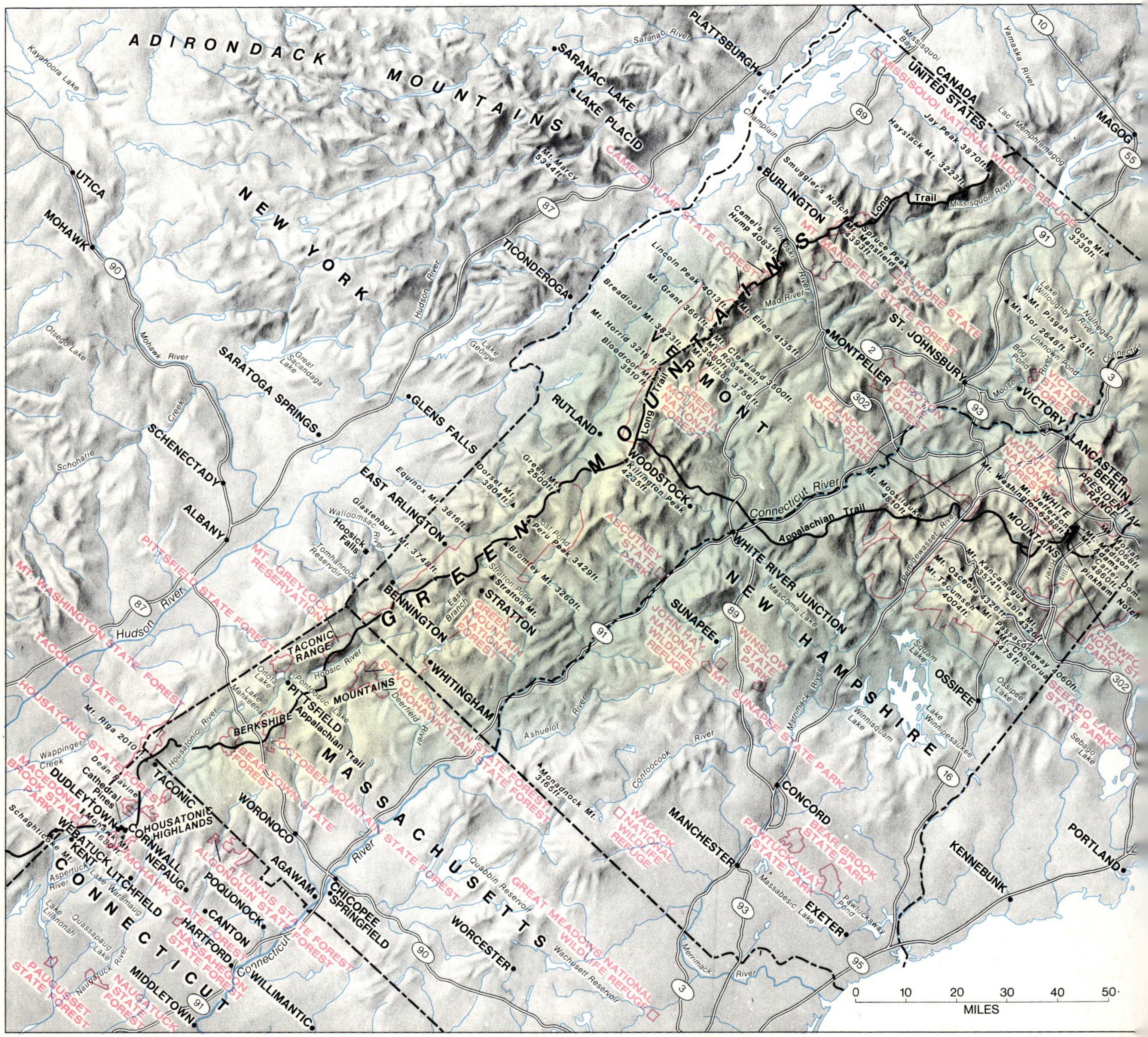

The chain of wooded highlands that runs through New England (shaded at right) is mapped in detail below. These uplands, the source of New England's major rivers (marked in blue), including the Allagash Wilderness Waterway (blue dots), have become much wilder than they were 100 years ago. Many of the hill farmers long ago gave up the unequal struggle against poor soil—a legacy from the region's burden of glacial debris. The public lands of the area (outlined in red) are linked by the northern 703 miles of the Appalachian Trail, ending at Mount Katahdin in Baxter State Park, and by Vermont's Long Trail. Both routes are marked with solid lines. The New England portion of the Appalachian Trail winds through the Green and White mountains, including some of the loveliest terrain in America.

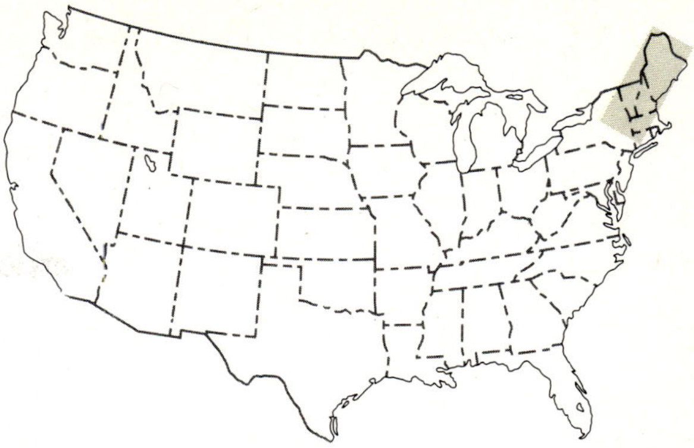

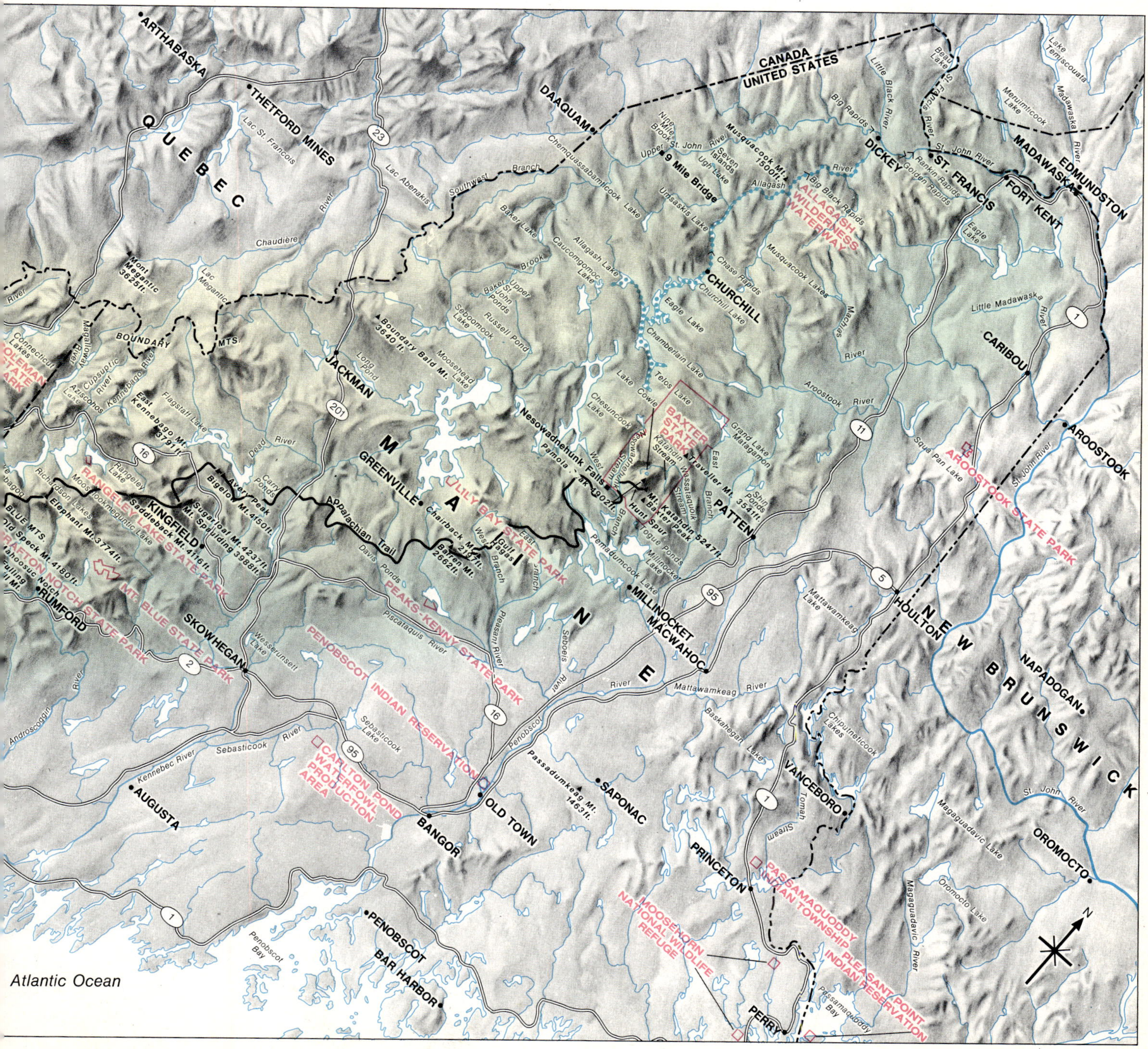

1/ Stone Walls and White Pines

Here further up the mountain slope/Than there was ever any hope/My father built, enclosed a spring/Strung chains of wall round everything/...The mountain pushed us off her knees/ And now her lap is full of trees. ROBERT FROST/ *THE BIRTHPLACE*

Yes, there is wilderness in New England. Even in this long-settled and crowded little corner of the United States, there are still wild and beautiful places left where you can climb a mountain, canoe a sparkling river or watch the sun set on a lonely lake. They are up there beyond the golden arches of McDonald's and the Bide-A-Wee Motel, hidden in between the road lines on the map. All you have to do is go and look. But hurry. And when you find a special place you learn to love, please keep it to yourself. They've already carved up the old place into too many quarter-acre plots.

There are, in fact, thousands of wildernesses scattered over New England, most of them in the high country that stretches north and east from the Berkshire and Taconic ranges of Connecticut and Massachusetts, through the Green Mountains of Vermont and the White Mountains of New Hampshire to the deep woods along the roof of Maine. Some of them, including the region's better-known peaks and public forests, are linked together by another line on the road map, the thin, dotted one that marks the Appalachian Trail. More and more people are leaving the roads to see New England that way, hiking different parts of the trail and its offshoots on weekends or vacations, or even doing the whole 703-mile New England section in one summer. But there are many other treasures in New England's attic, and even if you are an ardent hiker you cannot possibly see all of them along this path alone. If you would begin to know New England you must go back to it again and again, to many places and at many times of year. For it is an old region, rich in both natural and human history, and it does not reveal itself all at once.

You have to start somewhere, of course. Tucked away in the northwestern part of Connecticut, not too far from where I live, is a place as good as any, a little half-hidden, half-wild world I like to go back to from time to time. It is no Grand Canyon or Yosemite, that is certain. It is a softly rolling patch of the Litchfield Hills, about four to five miles long and less than half as wide, traversed by a segment of the Appalachian Trail as it meanders into New England from New York. But within these modest limits, a pleasant afternoon's walk, it offers a remarkable series of vignettes—a sampler of the region. First there is a darkly forested ravine with tall hemlocks, huge boulders and lovely waterfalls, a timeless, typical New England treasure that in itself would make the trip worthwhile. But there is also the eerie surprise and sadness of a genuine ghost town, its stone walls and old house foundations strangely buried in the woods. Then, from a low mountaintop, there is a classic view down to a pastoral Yankee valley and a tiny, whitewashed, steepled town. Finally, across the valley, there is a grove of magnificent, towering white pines. Here, almost as if you had placed it under a magnifying glass, is a snippet of New England's very fabric—a rich and intricate tapestry indeed.

The portal to this little world is called Dark Entry Road. Off the highway, after you pass a couple of houses and driveways, the steep blacktop ends and you enter the woods on foot. In the watery half-light of an early-winter afternoon, the leaves, and the hikers, have gone; the bones of the countryside, its trees and rocks, are etched unmistakably against the sky and a dusting of snow. It is one of the best times to go walking in a New England woodland, a bare and silent season in between the others, when you can begin to sense the essence of a place. The flush of spring, the fullness of summer, the blaze of autumn are left behind; the forest of maple, beech and birch, now stripped, is settling into winter. Only a few of the fallen leaves in the rustling brown carpet underfoot are still tinged yellow; only a few tiny red fruits cling to partridgeberry vines, passed over by the birds. The pulse of life is slowing: a single black-capped chickadee flits from branch to branch; a gray squirrel ripples up the trunk of an old gray birch and disappears into a hole; in a clearing a blackish porcupine with white-tipped quills dozes in a motionless ball, making the most of a weak ray of sun. It is a

A tangle of trees engulfs the stone walls of William Tanner's cellar hole in the abandoned hilltop community of Dudleytown, Connecticut.

somber, pensive kind of landscape, painted in black and white, browns and grays, touched here and there with the dark evergreen of feathery hemlock boughs and the shinier leaves of mountain laurel clustered above outcrops of gray-black ledge.

In this skeletal scene another element gradually reveals itself: the fact that people once lived here, in a settlement called Dudleytown. The path you are climbing was originally cleared not for holiday hikers but for 18th Century oxcarts and working horses to haul supplies and harvested crops. A depression you glimpse off to the left on closer inspection turns out to be not a natural rock formation but a man-made cellar hole, half filled with leaves and dirt. Two sides of the hole are still standing, built of flattish, mossy-green stones that were patiently fitted together without mortar by some early settler; the house, abandoned, has long since fallen in on itself and rotted away. In its place, rising through the spaces where a family once lived, are several maple saplings and two large and very silent beech trees.

Follow the path down to Bonney Brook, then up the dark ravine to Marcella Falls, a sparkling staircase of water that is now, in winter, a frosted wedding cake; at the top of it you can make out the mouth of a small stone dam, its sluiceway mutely open and its wooden mill wheel gone. Farther up the path, so rock-ribbed and precarious now you wonder how even oxen could have made it, is a gaping stone wall and another cellar hole; these mark the home of one of several Dudley families after whom the hamlet was named. Beyond is a little crossroads, quite deserted in the forest, then the ghostly sites of a school and more houses scattered through the trees.

Just before the path starts to dip back down into the valley, there is one last cellar hole, the roots of a great dead cherry-birch tree still grasping its stones. This place has a special fascination, at least for me: it was the home of a William Tanner, nicknamed Great Tanner, town records say, because of his unusual size. It was also the site of the first death that started the "Dudleytown Curse." One story has it that a neighbor was mysteriously murdered here in 1792; another says he toppled from the rooftree during the barn-raising, whether as a result of a loose timber or too much hard cider is not clear. Since then all sorts of dire deaths by lightning, fire and suicide have been ascribed to people who supposedly once lived in Dudleytown, or Owlsbury as it became known after the last human resident gave up and moved away in the early 1900s. Cousin Will, it seems, managed to live on to the age of 104, although he wasn't quite right in the head when he was finally carried out feet first to the village cemetery down below. But then, they said, anyone who could have spent that long scratching out a living on the mountain must have been crazy in the first place.

Crazy, perhaps, or just Yankee stubborn, or a little bit of both. For, apart from local legend, much of the hardscrabble history of New England is summed up in Connecticut's Dudleytown, and in a surprising number of ghostly counterparts scattered over backcountry hillsides through Massachusetts, Vermont, New Hampshire and Maine. Why were they abandoned? How did the stone walls and foundations get buried so deep in the woods? The story is as much a part of the New England landscape as its rocks and trees.

Three centuries ago, when land for farming was already becoming scarce along the seaboard, the early settlers struck out for the interior, where the bottomland was often so tangled and swampy that many took to the hills. There were wolves and Indians and an occasional mountain lion to worry about, but the main enemy was the land itself. Except for river meadows and a few clearings made by lightning fires or storms, most of New England was a dense, rocky, seemingly endless forest—a "hideous, howling wilderness," as one Boston minister rather peevishly observed during a trip inland in 1694.

The most urgent task was to carve back this forest to make room for corn and wheat, vegetables and apples, cattle and sheep. And so the great trees came down under the ceaseless blows of the ax, and were turned into firewood, houses, furniture, wagons, masts for ships and handles for more axes. But even when the stumps had been pried up by sweating men and animals, the soil was no bargain. One early Connecticut resident, Ezra Stiles, summed it up in a wry couplet: "Nature out of her boundless store/threw rocks together and did no more." What Stiles did not know then, of course, was that New England's soil had been scraped away by several great waves of Canadian ice, the last of which had retreated not too many thousands of years before. The little soil that had subsequently developed from erosion was mixed with an endless supply of stone glacial garbage that had to be levered loose, hauled off on stoneboats and piled to one side in walls. How many of these walls there are all over New England is anyone's guess; they are the region's well-known, well-loved trademark, stitching the countryside together almost everywhere you look. There is, in fact, only one more persistent feature of New England: its trees. For, despite the concerted war waged on them for three centuries by farmers, loggers, potash makers, shipbuilders, charcoal burners, railroaders and developers, almost everywhere in this well-watered, forgiving climate the trees have slowly returned. And even now they keep coming back, to overwhelm the farms, the fields, the burned-over hills and even the stones themselves.

This persistence is illustrated by two rather astonishing facts. By the mid-19th Century states like Connecticut and Vermont, originally 95 per cent or more forested, had been three-quarters cleared for farms, leaving only a quarter of the original trees. Today in these states the proportion is almost exactly reversed: forest covers three quarters of the land. In New Hampshire and Maine, never as widely farmed, the trees have reclaimed 85 per cent or more of their original domain.

It was the hill farms that started the cycle from cultivation back to wildness. The real curse of a Dudleytown was that raising anything on its high, thin soil was an almost impossible task. The growing season was not only short, but the land was shaded by surrounding hills much of the day thus apple trees grew dwarfed and corn and wheat were windblown and spindly; to make it worse, one-crop farming soon depleted the minerals in the soil. As the hillsides were stripped of more and more trees to increase the meager yield of crops—and to feed logs and bark into the growing number of sawmills, tanneries and iron furnaces that began to dot the landscape—the loose soil was washed down into the valleys, making the high ground still poorer. After a succession of backbreaking summers, and long, cold, lonely winters in the hills, it was little wonder that many families pulled up stakes and moved on. Farm abandonment, in fact, started in New England almost as soon as farming itself, as the hunger for better land pushed the settlers in ever-widening circles from the sea. The movement picked up speed in 1825 with the opening of the Erie Canal and the access it provided to the deep, level, fertile soils of the Middle West; migration accelerated again with the expansion of the railroads, and became a virtual stampede after the Civil War. Many who did not join it gave up their farms and went to work instead in the water-powered industries that had been developing along New England's rivers. The size of the exodus can be measured by the fact that in Vermont—a state that still numbers fewer than 500,000 people—200,000 left between 1776 and 1861 alone.

What happened to the abandoned farms and pastures can be seen repeating itself on many a hillside as you drive through New England today, in almost every stage of its succession. Where a farmer has recently left a field unworked, short-lived plants like dandelions and ragweed take over first, then grasses follow; gradually the grasses are crowded out by perennial weeds like aster, goldenrod and black-eyed Susan. Then, almost imperceptibly, an advance guard of shrubs and sun-loving "pioneer" trees moves in, shading out the weeds. In southern New England these are most often seen as the little regiments of red cedars that dot the hillsides, making dark exclamation points against the green and yellow fields. In northern areas, and at higher altitudes elsewhere, spruce or balsam fir is usually the first invader of cleared land; white birch, poplar and pin cherry, too, will sprout, particularly where fire has made a swath. But in the main, central part of New England, it is the white pine that most often moves in where the farmer has moved on. Sometimes it forms tight, pure stands of trees that over the years grow root to root, trunk to trunk, tall and straight in their mutual reaching for the sun.

It is just such a stand—described by some foresters as the finest remaining in the Northeast—that we see at the end of our afternoon's walk in the Litchfield Hills. The trees are indeed impressive: many are over 125 feet tall, three feet through the trunk, 200 or more years old. When you stand among them, the only sound is the steady, haunting wash of the wind through the treetops far above. The very air is primeval, dark, with a faint fragrance of pine needles. Perhaps inevitably, the grove has been named Cathedral Pines, and at least one pair of hikers who has passed through it has gone back to be married there.

The early settlers held the white pine in awe too, but in a somewhat different way, and this may explain why even these old pines, large as they are, are almost certainly not virgin, but second growth. Coming from an England that had been largely denuded of its great forests by their ancestors, the colonists could hardly believe the size and extent of the pine woods that confronted them, marching up the hills and river valleys as far as the eye could see. The northwest corner of Connecticut, where the pines reached farthest south into the deciduous forests of maples and beeches, hickories and oaks, soon became known as the Greenwoods for its abundant stands of these giants. Ezra Stiles measured one that was 15 feet around, and on the stump of another he counted 400 annual rings. Farther north, in New Hampshire, on the site of what is now Dartmouth College, stood a white pine that was judged to be 240 feet tall, and another, in Lancaster, New Hampshire, was recorded at 264 feet.

Near the Appalachian Trail in Connecticut's

Housatonic State Forest, a stream swirls through one of the waterfalls of Dean Ravine, its bed cut from ancient, tilted layers of schist.

The awesome sight of trees like these, stretching into northernmost New England, was not lost on Captain George Weymouth of England's Royal Navy, who explored the coast and river mouths of Maine as early as 1605; he quickly saw the potential in the tall, straight trunks for sorely needed masts for the growing English fleet. No wood so light and strong could be found in Europe in such great lengths; the English had to splice their best masts out of pieces of Riga fir imported from the Baltic—a supply that could easily be cut off by war. Before long the pine was being carried to England in special "mast ships," built with large stern ports for loading timbers over 100 feet long. At the same time the white pine was also becoming a mainstay of New England's domestic economy, its basic tree for home consumption as well as its most important export along with fish and furs. Most treasured of the varieties of white pine was the so-called pumpkin pine, a slow-growing product of the virgin forests whose wood was not only light and strong but pumpkin-gold inside, and so soft and smooth it could be worked with almost unbelievable ease. It was this wood that went to make houses, furniture and wide-board paneling, as well as the exquisite moldings and fanlights of the finest New England mansions, and the elaborately carved figureheads of the merchants' and whalers' ships.

By 1691 the demand for white pine was becoming so great that the Crown began to reserve the best mast trees—all those over 24 inches in diameter and located within three miles of water—by notching into their trunks an upward-pointing brand called the King's Broad Arrow. This may have been a tolerable royal prerogative while the supply seemed unlimited, but well before the time of the Revolution the broad arrow had become a symbol as hated as the Stamp Act and the tax on tea, and in the backwoods of New England this may have been an even more telling factor in bringing on the war. To the woodsman these trees were as much his for the taking as the king's. And so, behind the backs of the royal "mast agents," sent over to enforce the policy, many Yankees chopped the trees down, effaced the broad-arrow brand and floated the huge logs downriver for sale to the highest bidder. When a spy system was set up to catch violators, some woodsmen disguised themselves as Indians—years before the Boston Tea Party—and either beat up the king's spies or cut the trees at night; in retaliation the British burned the loggers' sawmills and drove them from their homes. It is hardly a coincidence that at the Battle of Bunker Hill the Americans fought under a flag that bore a pine tree in one corner. Massachusetts adopted a naval ensign with a pine tree above the legend "Appeal to

Heaven." And Maine, when it later broke away from Massachusetts, christened itself the Pine Tree State.

After the war was over, the relentless pursuit of the pines went on. In a memoir of his early logging days in Maine, *Forest Life and Forest Trees,* John Springer called the giant pine that rose above the spruces and firs "the whale of the forest." And like the whales New Englanders were pursuing around the world, the pines were eventually hunted almost to extinction before major stands were given a chance to come back under proper forest management. The pines stood in groves, most numerous in central New England but snaking up the lakes and rivers northward wherever temperature ranges were not too extreme and there was well-watered and well-drained sandy soil. The early "timber lookers" usually chose a towering pine to climb—after lodging a cut spruce against it as a ladder to reach the lowest branches—in order to spot more stands of pine on the horizon. "Such views," wrote Springer, "fill the bosom of timber-hunters with an intense interest. They are the object of his search, his treasure, his El Dorado, and they are beheld with peculiar and thrilling emotions."

By 1850, when Springer wrote, most of the virgin white pine in New England had already fallen to the ax and loggers were moving west in search of more—to make more houses on the prairies, more covered bridges, more masts and spars for clipper ships. When the second growth of "old-field pines" that had sprouted on the earliest cut-over hills and pastures reached marketable size, these too were cut down, in a second great harvest between 1875 and 1925. But by then the lessons of headlong exploitation were beginning to sink in: the forest, it appeared, was not endless after all. Once more the white pine served as a symbol, if now a tragic one: what had happened to it in New England helped to focus the battle for timber conservation and establish the national parks and forests in both the East and West.

Many New Englanders were in the forefront of the movement, pointing out the need for sensible forest management to protect watersheds, wildlife and recreational values, as well as to ensure a more reliable supply of the trees themselves. One was George Perkins Marsh of Vermont, whose writings were a wellspring of the conservation movement in America *(Chapter 2).* Another was Percival Proctor Baxter of Maine, who rescued the Mount Katahdin area and turned it into what is perhaps New England's finest preserve *(Chapter 6).* Less celebrated are hundreds of other conservationists who have worked, and still work, to save the patchwork of New England's wilds. Their names are not well known—a botanist here, an Audubon Society member there, an editor or farmer or ordinary citizen somewhere else. But they have been, and remain, the key to the region's unique and changing landscape: unlike the vast national parks and forests of the West, which were largely created out of virgin territory by broad strokes of a governmental pen, most of New England's wild places have had to be painstakingly reclaimed by countless individuals out of old and often well-used scraps of land. These precious scraps, pieced together, resemble nothing so much as an old New England heirloom quilt that is made, frugally and lovingly, to be passed down from one generation to the next.

The process goes on today, with committees and computers suited to the temper of the times. In an attempt to coordinate the myriad conservation efforts, a New England Natural Resources Center in Boston is identifying, evaluating and cataloguing in a data bank facts on nearly 5,000 natural areas scattered throughout the six New England states, and is seeking means and funds for preserving those whose future is in doubt. The fragments range from a bed of rare ferns in New Hampshire to an Indian burial ground in Vermont, from a waterfall in Massachusetts to an 8,000-acre lake, named Umbagog, that straddles the border between New Hampshire and Maine. What makes the task doubly complex is the fact that two thirds of the valued areas are still in private hands; if they are to be permanently preserved in their natural condition some will have to be bought by one state agency or another, while others will have to be set aside by means of easements granted by owners, by trusts or by outright gifts.

One pattern was set nearly a hundred years ago by a man named R. J. F. Calhoun, a Connecticut dairy farmer who had grown up with and loved Cathedral Pines. In 1883 he could see the white-pine loggers coming once again, and he decided to buy the 42 acres the trees stood on to save them from being cut and sold. His heirs have kept their piece of the patchwork in the family until recently, when they turned it over to the Nature Conservancy to place it in permanent public trust.

Besides their size and majesty, there are two other things that you notice when you stand among these trees. One is that they seem to be as tenacious as they are tall: one giant specimen appears to be growing right out of a huge boulder, elevated as if on a platform a good eight feet above the ground. On one side the tree seems to have only a tentative hold on the rock; on the other, however, you can see two massive trunk roots winding like tentacles down the stone to anchor themselves in the soil. At some point two centuries or more ago, when the land was cleared by fire or logging, a little pine seed must have drifted into a crevice in this rock, where disintegrating rock particles and a few rotting leaves had helped start a little pocket of soil. Nourished by this tiny seedbed, the young tree had slowly sent exploring roots in search of more permanent sustenance and stability, right down the granite face, until it had at last gripped the rock in its embrace.

The other thing you notice among the huge rotting trunks of older fallen trees are more seedlings and young trees. They are not white pines, which cannot survive in their parents' dense shade, but a new generation of hemlock, maple and beech, more shade-tolerant species whose seeds have been brought into the grove by wind, animals or birds. They are biding their time, growing slowly but steadily as they take advantage of a few patches of light that filter in where the older trees have thinned or fallen. In many places some of the older hemlocks have already grown almost as big as the pines themselves. Someday they will take over completely, for the field-to-forest cycle, though more than two centuries old, is not finished yet; the next forest here will be quite different from the one we see today, and the white pine will have to move on to colonize some other sunny place.

For a good many years to come, however, barring a disastrous fire or windstorm, people will be able to come and marvel at this grove. And that is good, because New Englanders like to cling to their history, and their wild places, even as the tree clings to the rock.

A Winter Calendar

PHOTOGRAPHS BY SONJA BULLATY AND ANGELO LOMEO

New Englanders often grumble wryly about their weather ("nine months of winter and three months of damn poor sleddin'," goes one old saying). But despite all the talk of stalled cars and frozen pipes, most of them would not have it otherwise, for while New England winters can be bleak they can also be full of beauty, variety and surprise.

Part of the weather's appeal is its very unpredictability: almost every day is different from the one that went before. A blizzard may march in unexpectedly, turning the hills and valleys blurred and gray with stinging snow. Then, just as abruptly, morning will bring a landscape of dazzling whiteness under unbelievably blue skies. Even the type of storm is not easy to predict: one may be so violent it cuts visibility to zero and cracks ice-laden branches off the trees; the next may be the kind of gentle emptying of the skies in large, soft flakes that moved Robert Frost to write so eloquently of watching the "woods fill up with snow."

In the calendar of winter, each month has its mood, suggested here in photographs of the Green Mountains of Vermont. Not long after the leaves have fallen, the mountains get their first dusting of white, their contours still visible beneath the fine tracery of naked branches and the dark tufts of evergreens. As the season progresses, hillsides are blanketed in deeper drifts and trees are turned into statues by coatings of snow and ice. The amount of snow varies, with true New England individuality, from year to year and place to place. In some winters the land remains strangely bare for two or three weeks at a time. In others, snow has piled deep enough to topple locomotives trying to plow out railroad track. The most celebrated year of all was 1816, still referred to as "Eighteen-hundred-and-froze-to-death," when winter never stopped; farmers shoveled themselves out in spring only to see their crops blackened under June and July snowstorms and heavy August frosts.

But in most years the land starts to unlock itself from winter's grip by March or April; the sound of mournful winds gives way to the soft plop of snow falling from branches and the crack of ice breaking up on the streams. Young ferns and flowers begin to show through melting patches of snow, and the leaves of birch and beech unfurl. New England's winter, actually four or five months long, is giving way to spring.

A sugaring of early snow lies on a still-golden maple leaf in late October. Soon the fallen leaves will turn a muddy brown, and the last reminders of a flaming autumn will be smothered under a thick mantle of white.

A November blizzard whistles through a hillside forest of birch, beech and maple dotted with dark-green spruce. On the ground, patches beneath trees and in damp spots are the last to be covered by the enveloping snow.

Just after a December storm, a late-afternoon sun glows mistily through a clearing cloud bank, picking out the bare branches of trees decked in snow and touching the white hillside with the gold of a winter sunset.

In late December, winter has set in hard and gray in Smuggler's Notch below Vermont's Mount Mansfield; rime-whitened trees stand in stark

contrast to icicle-laden outcrops of rock. Snow covers a cascade of boulders loosened from the mountain's walls by endless freezing and thawing.

By January, snow on the exposed face of Mount Mansfield (left) has drifted to a depth of 10 feet. Like most drifts formed in the lee of objects, this one has an oval depression scooped out by winds eddying around the rock.

By February, the land seems locked in winter forever. The black branches of trees are hidden in a casing of white ice, and water spilling over a rocky outcrop has frozen, layer by layer, to form the wall of an icicle cave.

Fir trees, plastered with wet, driving snow that has been frozen and carved by wind, stand like the sculptures of a madman on Mount Mansfield's summit in March. Occasionally a tree cannot take the strain, and snaps suddenly, with a crack like a pistol shot that echoes through the mountain stillness.

April is the season New Englanders call unlocking, when the long winter at last begins to loosen its grasp. With it come the new yellow-green shoots of falsehellebore, which poke up here through a collar of evergreen ferns.

A brief storm in May frosts the ground and hemlock branches one last time with snow, but birch and maple in this glade have already sprouted new leaves, and spring flowers will soon shoot up through the forest floor.

2/ The Mountain Roots

We have met here to . . . consecrate this extensive wilderness and give it a new name, which new name is Verd-Mont in token that her mountains and hills shall be ever green and shall never die. REVEREND SAMUEL PETERS/OCTOBER, 1768

There is an old saw to the effect that if you flattened all the wrinkles out of Vermont it would be as big as Texas. With apologies to Texas, I cannot imagine a more horrendous fate. For, like almost everyone who has ever spent some time there, I love the rumpled carpet of the Green Mountains, with the villages nestled in its folds. I only wish people would leave it alone.

I sometimes wonder why Vermont, a little keystone-shaped state dropped into the top of New England alongside New Hampshire, should arouse quite the deep passions that it does. I suppose it is the ultimate Yankee pastoral, a land of milk and maple syrup, weathered barns and covered bridges and bean suppers at the village church. Some years ago a Kansas lawyer actually proposed that the whole state be made into a national park so the rest of the country could go there to see how their great-grandparents lived. This probably amused quite a few Vermonters, who have become more or less accustomed to the silly things outsiders say from time to time. Nevertheless, the lawyer had a point, and one that keeps coming up with increasing frequency among admirers of the state. Will Success (condominiums, leisure communities, pizza parlors, snowmobiles) Spoil Vermont?

Take a look at Vermont itself. It is not only small, the 43rd in size among the states and the least populous one in New England, with about 460,000 permanent residents; it is also peculiarly unflat: a virtually continuous, border-to-border landscape of hummocks and peaks, rolling up from the Connecticut River Valley on the east and down again to the Lake Champlain and Hudson Valley lowlands on the west. In all there are probably over a thousand bumps that rise above 2,000 feet; all are loosely referred to as the Green Mountains, though technically the name is reserved by geologists for the band of higher mountains that form the apex, the north-south backbone of the state. This major wrinkle at the western edge of New England actually starts in Connecticut's Litchfield Hills, extends north as the Berkshires in Massachusetts, becomes a broad upland in the Green Mountain National Forest in southern Vermont, then marches north of Rutland to form Vermont's Presidential Range—Mounts Wilson, Roosevelt, Cleveland, Lincoln, Grant. Just south of the Winooski River, which cuts through the range's backbone on its way to Lake Champlain, is the 4,083-foot peak called Camel's Hump, the favorite of many Vermonters, a wind-scoured, slightly lopsided summit that very much resembles its name. North of the Winooski Valley the ridgeline ascends once more to the treeless top of Mount Mansfield, the state's highest peak at 4,393 feet, then trails off to climb a few last major summits, like Jay Peak, before subsiding across the border in Canada. In the early days there was some doubt as to whether Camel's Rump, as it was then called, or Mansfield was actually higher; the matter was settled with true Vermont pragmatism when a hunter climbed to the top of Mansfield, placed a small rifle ball in his fowling piece and pointed it at the top of the rump. The ball rolled out.

I first started going to these mountains when I was in college, mostly to ski at places like Mansfield near Stowe, which was already being billed as the "ski capital of the East," as well as at Bromley Mountain and Woodstock farther south. Mansfield was the most spectacular, and the best skiing, particularly the old Nose Dive trail with its hair-raising Seven Turns. But Woodstock always seemed more like Vermont. It had just one steep little hill outside of town called, rather grandly, Suicide Six; ascending it, you had to clamp down hard on the icy, fast-dragging tow rope if you wanted to make it all the way to the top. The builders of Suicide took some pride in the fact that they had practically invented the American ski area, having rigged the first ski tow in the country by hooking a line and flywheel to an old Buick motor (this also came in handy for *uphill* jumping, a sport not practiced much anymore). That was back in the baggy-knicker days of 1934.

Since then a great many more ski areas have been built in Vermont

—too many, some Vermonters say. At last count there were 42 of them scratched into the mountainsides up and down the state. Quite a few, like proper modern ski resorts, are equipped with cozy gondola lifts, fancy base lodges, heated swimming pools, rock bands and small cities of more-or-less-Swiss chalets. When attendance rises over so many thousand on a busy weekend, however, there is grumbling in the lines of people waiting for the lifts. One result is that quite a few of them have taken to the woods—on snowshoes, cross-country skis and even on foot. The woods in winter are still a beautiful and relatively uncrowded place, and a much better way to see Vermont.

This means seeing the Green Mountains, for as someone succinctly put it, the Green Mountains *are* Vermont. You can enjoy them from any one of a number of high or low vantage points, rising in all directions in gentle waves and swells. In summer they are indeed green—a dark, rich, gorgeous green. In fall they are painted in brilliant reds and yellows, before turning a striking russet and black, etched with the bright slashes of birch trunks. In winter they are dazzling white, and then in spring they sprout a fresh palette of pale greens and yellows before they mellow into their deep, rich green again. At no season are even the highest of the peaks raw or overpowering, like some of their brash young cousins in the West. They are eminently old and rounded, as New England mountains should be, big enough to command respect, even stern and distant in their wintry moods, yet in essence of an almost human scale. It is these mountains that provide the foundations of the state's economy, not only the ski business and the fall colors and the tourists that bustle about in the antique shops, but also just about everything else on which Vermont depends: trees for its many wood industries; food and cover for its wildlife; acreage for upland farms and pastures; marble, granite, asbestos and a good many other dividends. Not the least among these is a steady, year-round supply of water, which makes for exceptional fishing and some good canoeing, and without which nothing much else could exist.

It is not surprising that these mountains have a hold on people who live among them, that their stony soil and long winters, as well as their beauty, have left their marks. (Where else can you find both a Delectable Mountain and Mount Horrid—or, for that matter, a Ticklenaked Pond?) The mountains seem to bring people together in common bonds, and yet at the same time keep them judiciously apart, channeling their comings and goings into the small valley settlements whose houses, barns

To most people New England's trademark is granite, but the region abounds in a variety of other rocks as well, as notable for their hues and patterns as they are for their great age. On these and the following two pages, arranged in order of seniority, are banded limestones and other sedimentary rocks; black marble streaked with quartz; once-molten pegmatites studded with tourmaline; and such lesser-known rocks as phyllite, crushed into fluid diagrams by the folding of the earth's crust.

GNEISS, VERMONT (1,100 MILLION YEARS OLD)

DEFORMED SLATE, MASSACHUSETTS (570 MILLION YEARS OLD)

MARBLE, VERMONT (550 MILLION YEARS OLD)

LIMESTONE, MASSACHUSETTS (500 MILLION YEARS OLD)

PHYLLITE, VERMONT (480 MILLION YEARS OLD)

SLATE, VERMONT (475 MILLION YEARS OLD)

PEGMATITE WITH TOURMALINE, MAINE (360 MILLION YEARS OLD)

PEGMATITE WITH QUARTZ, MAINE (185 MILLION YEARS OLD)

SCHIST, VERMONT (470 MILLION YEARS OLD)

SANDSTONE, MAINE (400 MILLION YEARS OLD)

ARKOSE, CONNECTICUT (185 MILLION YEARS OLD)

GRANITE, NEW HAMPSHIRE (185 MILLION YEARS OLD)

and white steeples fit so magically into the contours of the land. ("No Americans," author Stewart Holbrook once wrote, "have fitted the architecture of their villages, their homes, their highways, even their meadows and pastures, to the scene with a better eye to esthetics as well as convenience.") Even the state's name reflects a fine eye for design. "Voilà! Les monts verts!" the French explorer Champlain is said to have exclaimed in the summer of 1609, gazing across to the emerald slopes from the lake named after him. When Vermont's founding fathers met in 1777 to draw up a constitution, some wanted to name their new home New Connecticut, recalling their older homes down south. Fortunately, wiser counsels prevailed.

One's regard for this beautifully wrinkled landscape only grows with a knowledge of how astonishingly old it is. Much of the Green Mountains' intricate geologic history, like that of the rest of New England, is so fragmented, concealed or simply vanished that for long it remained a mystery, and even today some of the fine points are still being discussed. Oldtimers talk of "the glacier," somewhat wryly because even after several generations on the same farm they still get a fresh crop of rocks every year, heaved up by the frost. Their grandfathers, in turn, probably talked instead about "the flood"; for until the pioneering Swiss geologist Louis Agassiz came along with his glaciation theory in 1837, most people thought the boulders could only have been washed in by some great liquid deluge long ago.

What did happen goes back far beyond any flood or ice age, however, to what geologists call the billion-year-old basement, of Precambrian rock. This ancient terrain of gneiss was submerged beneath a succession of inland seas that once covered much of New England. Rivers pouring into these seas laid down layer upon layer of sediment, which solidified under the tremendous pressure of the layers above them and formed sandstones and limestones.

About 450 million years ago, as the slowly drifting continents of Africa and North America were making their journey towards each other, the layers and the underlying basement were caught in the great continental vise, heaved up, squeezed and recrystallized into different kinds of rock—marble, shale and schist. This was the so-called Taconic Revolution, in which great waves of rock actually crested and broke, sliding off the top of the Green Mountains and Berkshires to form the Taconic Mountains a few miles to the west. The Green Mountains, or what is left of them, are the oldest mountains in New England,

and are about seven times as old as the Rockies, which were erected a mere 60 million years ago. When the ground stopped shaking after their birth, the Green Mountains may have stood as high as the Himalayas do today. There were subsequent periods of deformation and uplift; the last of these occurred about 200 million years ago, when Africa and North America finally parted company, cracking and buckling the mountains still further as the land masses opened to form the Atlantic Ocean. Since that time, erosion has stripped away as much as six miles of rock, and what we see today are only the stumps, or roots, of this once-mighty range. The oldest rocks that lie exposed in New England —the billion-year-old basement—spring up in the central ridgeline of the Green Mountains near Stratton Mountain, taper to a point again in Northern Massachusetts near North Adams, then reappear to run down the Berkshire highlands into the Litchfield Hills.

Some graphic indications of the mountains' hidden, tortured past can be glimpsed along the highways, particularly the relatively new Interstate 89, which cuts across the north-central part of the Green Mountains from White River Junction on the Connecticut River, following the White and Winooski rivers to Burlington on Lake Champlain. Each time the road builders blasted through a placid, rounded hillside they opened up a new and more violent-looking cross section to view: brown shales, green serpentines, gray gneisses, white marbles —the strata often upended and sometimes even warped over backwards like bits of a crazily twisted layer cake.

Almost everywhere in the Green Mountains you can also see evidence of the last series of major geologic happenings that beset New England: the colossal waves of ice that advanced and retreated beginning a million years ago. At Lake Willoughby, north of St. Johnsbury, some of the surrounding peaks are streamlined on their north sides and plucked jagged on the south, indicating that the huge ice mass rose up and over them, grinding them smooth on one side and pulling rock loose on the other. If you look through the slot formed above the lake by its guardian peaks, Mount Pisgah and Mount Hor, you can see how a rough tongue of ice and rock licked and scraped the valley, transforming it from a V shape into a U, deepening it enough to form a basin for the lake. In many valleys you can also spot drumlins, rounded little hills created when clay and gravel traveling in the bottom of the ice were caught on protruding knobs of bedrock and formed up around them; many of these mounds have been cut into, in unsightly fashion, for their valuable road-building materials and fill. And, of course, there

The ice-fringed East Branch of the Deerfield River in Vermont's uplands runs swiftly in March, swollen by melting snow.

are all those loose stones, piled in walls or just lying around in the forests and fields. Vermont's biggest single glacial "erratic," or wandering boulder, looming in the woods near Whitingham in the southern part of the state, is a pebble 25 feet high, 40 feet long and 125 feet around. They call it the Green Mountain Giant.

Each time the ice advanced, the plants and animals retreated; each time the ice retreated, they in turn started slowly north to recolonize the ice-cleared, barren land. After the last great retreat of the ice cap began some 15,000 years ago, lichens once more gained a foothold on the rocks, helping erosion to break them down into nutritious soil. The changing climate became hospitable to larger plants like tundra grasses and sedges; these plants, in turn, were followed north by cold-tolerant shrubs and trees. The procession past any given point in New England can be charted by studying samples of the successive layers of pollen blown into bogs that had formed in depressions left by the melting ice. The oldest and deepest layers pulled up by a long coring pipe almost invariably yield pollen from hardy evergreens like spruce and fir, indicating the cool climate that immediately followed the glacial period. Subsequent layers show a preponderance of pine and hemlock, revealing the warming trend; above these are evidences of deciduous trees like maples and beeches, and finally types like hickory and oak, which thrive in still-greater warmth.

The most drastic change that the forest has undergone since the last ice age, however, started a mere three centuries ago with the settlers' logging and clearing for pastures and farms. In the late 18th and early 19th Centuries millions of acres of trees were cut and burned for potash, an early mainstay of the American export trade, used in making soap and glass. With the discovery in Europe of a cheap substitute—soda ash made from salt—the potash industry withered. Vermonters, casting around for a new economy, turned to raising sheep. They soon discovered that Spanish Merinos provided the finest wool, and imported a few hundred to start their flocks; the sheep took so well to their new home that at the height of the "Merino mania" there were 1.7 million of the animals in the state, outnumbering the human population 5 to 1. Eventually prize Merino breeding stock was sent from Vermont to the American West, South America and Australia; there, on the vast plains, the flocks proliferated to knock the bottom out of Vermont's own sheep-raising business. This turn of events was probably a godsend, for the close-cropping, sharp-hoofed animals might have nibbled and trampled the Green Mountains into a hopelessly eroded desert.

Around this time, at the onset of the Civil War, one farsighted Vermonter took a sobering look at the lay of the land. Compared to other conservationists, George Perkins Marsh is not particularly prominent in the history books today, but he could well be called America's first ecologist, and his ideas have had profound long-term results. Thoreau, Emerson and other intellectuals were already calling for a new attitude toward the wilderness on ethical or romantic grounds, but in *Man and Nature,* published in 1864, Marsh issued a solid economic warning against the effects of man's heavy hand. His main target was the indiscriminate lumbering then going on, which he felt to be in the long run "the most destructive among the many causes of the physical deterioration of the earth."

Born in Woodstock in 1801, the son of a well-to-do lawyer and landowner, Marsh had dabbled in virtually everything that challenged his unusually restless intellect: lumbering, sheep raising, construction, banking, teaching, law. Along the way he also found time to design Vermont's classic gold-domed, granite capitol in Montpelier, to produce a scientific analysis of the state's fisheries and to serve as a Congressman in Washington. But it was his reputation as a self-trained scholar—he was proficient in 20 languages, and among other things wrote the first Icelandic grammar in English—that apparently won him his appointment as Lincoln's ambassador to the new kingdom of Italy, where he served for more than 20 years until his death in 1882.

While in Europe, Marsh had ample chances to study at first hand the end results of deforestation, to which he closely linked the decline of the Roman Empire and other civilizations around the Mediterranean. He observed the continuing spectacle of depleted and eroded farmland, altered microclimates, disastrous landslides and floods, and he hailed the corrective reforestation and forest-management techniques that were slowly evolving in Germany and France. In his book, rooted in his native Vermont and written from a European perspective, he portrayed the forest as the world's great modifier. He described how trees checked high winds, anchored the soil, evened out extremes of temperature and moisture, broke the force of rainstorms, held and shaded the insulating winter snows, and gradually absorbed the precipitation into their great spongelike reservoir of humus, roots and permeable soil. Against this balanced forest organism Marsh contrasted the cleared, overgrazed, windswept hillsides, particularly in northern climates, where spring thaws caused water to run quickly over hard-frozen ground to swell and flood the ice-bound rivers below. Marsh—who also delighted in such subtler forest details as the role of earthworms in perforating and enriching soil, and that of birds in distributing seed—foresaw that properly preserved and managed forests could be "an inexhaustible and self-renewing supply" of a material increasingly indispensable to mankind.

Man and Nature, subtitled "Physical Geography as Modified by Human Action," was a surprising success, both in America and abroad, at least among the growing number of people beginning to worry about such things. It provided powerful arguments for practical preservationists like John Muir and Gifford Pinchot, and is often credited with inspiring the movement for national forests and parks. Historian Lewis Mumford called the book the "fountainhead of the American conservation movement"; Stewart Udall has since termed it the "beginning of land wisdom in this country."

I was thinking about Marsh and his book when I went into the Green Mountain National Forest one March day not too long ago. The principles he wrote about were indeed at work. In the valleys below the mountains the snow had vanished under the sun in an early spring thaw; the streams were boiling with cloudy blue meltwater that had run rapidly off the still partly frozen fields. Some lowland farms were already flooded, the water lapping around pieces of stranded farm machinery. The ski area operators had nothing to look up at now but tattered islands of snow in sloping seas of mud, and had sadly closed their ticket booths and gone home.

The road from East Arlington up into the forest was hung with a chain and one of those terse, hand-lettered signs one is apt to see at this time of year: "Closed—Mud." But circling the long way around Stratton Mountain I found enough pavement to get in the other way. Up here, at the modest altitude of 2,000 feet, the slightly cooler temperature, combined with the holding, shading action of the forest, had preserved almost a foot of pure white snow. A little smug that I was stretching the skiing season, I put on my cross-country skis and started up the Long Trail, which in southern Vermont incorporates the Appalachian Trail and at this point follows the bed of an old logging road. For a mile or more my only company was the East Branch of the Deerfield River, a pretty, sparkling black stream that gurgled along in a well-behaved manner, gradually absorbing other rivulets and streams as it went. In open places along the trail, the snow was sculptured in wind-blown drifts, here and there imprinted with the tracks of deer or snowshoe hare. The forest was marvelously quiet: except for an occasional chittering bird, the only sound was that of water, melting, trickling, running, splashing, gathering forces for its march to the sea. Here, as elsewhere, the mountains were not only the roots of older mountains, and economic and spiritual roots for people who live among them, but also the roots of the great branching river systems that supply water and power to towns and cities many miles away.

There were other signs of spring along the river: the sharp spears of beech buds starting to unfold, the sprouts of early spring flowers beginning to poke through bare patches in the snow. Climbing up the trail into the forest, which was now growing darker with spruce and fir, I passed a beaver pond, its drowned trees gray sticks standing like skeletons in the ice. On a rise, the dark woods suddenly glowed with a lovely pinkish-orange light, reflected from the trunks of a hundred big paper-birch trees. Finally, below me, lay Stratton Pond, one of the finest of Vermont's high lakes and a longtime favorite of hikers on the Long Trail. Someone else had come up to enjoy it, I discovered; along the shore were snowmobile tracks and beer cans scattered in the snow. I'm told that snowmobiles have helped a lot of people to get out of the house and away from television during those long, gray New England winters, that they bring families together. I also know that they shatter the forest's silence, forcing winter-weakened deer to run to exhaustion, and that they open up packed-snow avenues that house dogs happily follow, running into the woods to finish the deer off.

On the way back down the trail, I met other travelers in the woods, a bearded young man and two girls from Bennington, plodding stoically up the path on snowshoes, carrying orange and blue backpacks, sleeping bags and topographic maps. They were going to camp for a night or two at the pond, then climb Stratton Mountain and return by another trail. Knowing of the Long Trail's growing popularity, I asked them what it had been like up here last summer. "Grand Central Station," was the wry reply. "Winter is the best time of year."

By the time I got back down to the East Branch, the low clouds had broken; the sun was dancing on the water and on the delicate, lacy collars of ice that still rimmed its rocks. Beginning to sweat now in the early spring warmth, I peeled off knapsack, parka and skis, scooped up a drink and stopped awhile by the edge of the stream.

Marsh, I recalled, had suspected that forests not only acted as great sponges to regulate streams like this one, but that the trees actually at-

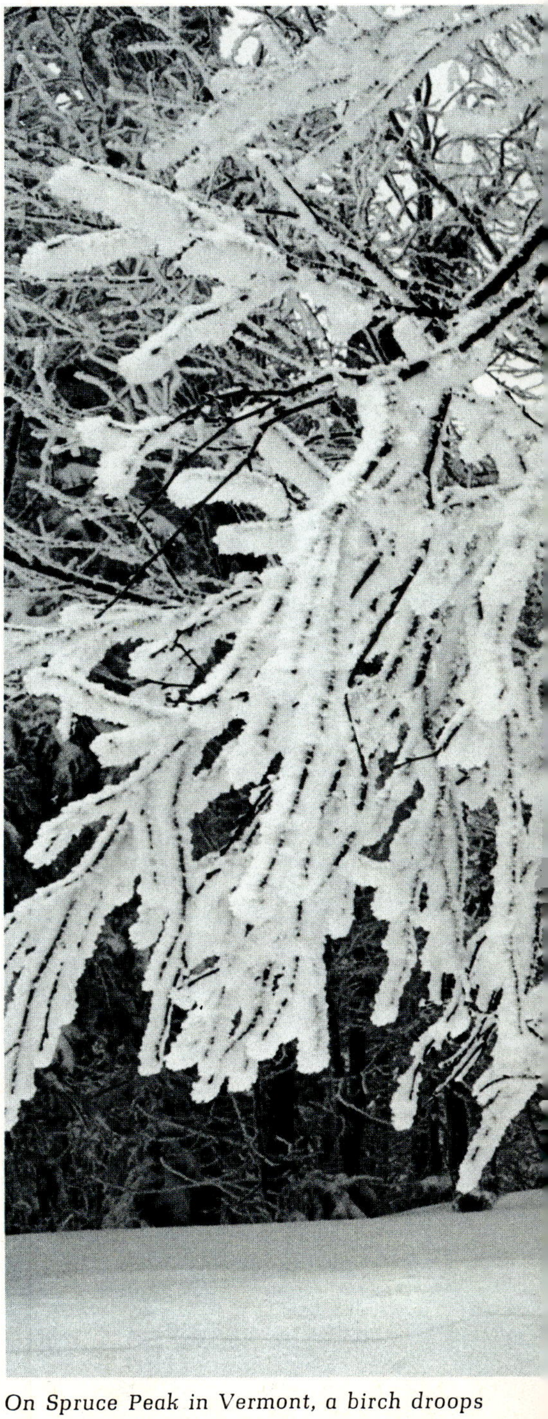

On Spruce Peak in Vermont, a birch droops

under an icy coat of fog droplets combed from the air—moisture that will augment the forest's ground-water supply at the first thaw.

tracted more moisture than open land. It was up to later scientists to prove him right—that wooded uplands literally comb great quantities of water out of the air and thus add significantly to the ground-water reservoir. One of these scientists, an adopted native son named Hubert Vogelmann who teaches botany and ecology at the University of Vermont, has spent a good deal of time in the Green Mountains with several associates measuring such natural phenomena. In one of their experiments, they set out pairs of moisture gauges in forest openings at various altitudes and came back at later intervals to observe results. The gauges were simply open-topped, one-quart oil cans fixed to the top of six-foot posts; one of each pair was left empty to measure straight rainfall, the other stuffed with a double roll of aluminum window screening to intercept moisture and precipitate it out of the air. Over a two-month period one summer, the gauges confirmed what the experimenters believed. Not only did rainfall increase with elevation but "fog drip" rose even more sharply: at the higher elevation the screen-equipped cans gathered two-thirds more water than the screenless ones. The fine mesh of the aluminum wire, like the fine mesh of branches, leaves and evergreen needles in the high forests, acted as a screen to remove the moisture from fog and low-lying clouds that frequently sweep across the upper mountain slopes. The tiny cloud droplets, too small to fall as rain, gathered on the needles and twigs, as they did on the aluminum mesh, eventually coalescing into drops heavy enough to run to the ground. The experiment also revealed another interesting fact: the moisture thus precipitated supplies the forest with water at the times it is most needed—cloud condensation contributed almost five times as much moisture as did rainfall during the dry days of late August.

Further studies of the mountain forest led Vogelmann and others not only to appreciate its importance as a water comb and storage sponge, but also to realize what a fragile mechanism it is, especially at higher elevations, where the greater total precipitation can combine with the steeper slopes and shallower soils to create serious problems if the natural cover is disturbed. The critical ecological "break" was pinpointed at about 2,500 feet, an easily visible line on most of New England's mountains because it is where the northern hardwoods like sugar maple, beech and yellow birch begin to give way to evergreens such as red spruce and balsam fir. Above this line the thin, acidic soils and harsher climate drastically limit the number of plant species that can survive. Those that do have only a marginal toe hold for their roots in the shallow soil, and are limited in their capacity to reestablish themselves once removed or disturbed—whether by logging, ski slopes, parking lots, access roads or clusters of vacation houses. Road construction and building foundations are especially detrimental to natural water drainage, and septic systems can add a fresh burden of water and waste-matter buildup that the ground simply cannot absorb.

One result of such studies, and of a growing general concern about overdevelopment of the state, was the passage in 1970 of Vermont's tough and far-reaching Land Use and Development Act 250. Among its chief provisions, Act 250 prohibits projects that adversely affect scenic values or threaten irreplaceable natural areas; it particularly restricts building above 2,500 feet unless stringent criteria are met to preserve the existing plant cover and the quality of the water. The legislation has already forced the installation of costly sewage processing plants at expanding resort areas, as well as the departure of some out-of-state developers who had been holding large tracts of land. Act 250 also led to the drafting of an overall land-use plan to guide Vermont's future growth—one of the few of such sorely needed statewide plans in the United States. At the core of the new laws is the belief that the highest and best land use of the Green Mountains, or any mountains, is as a source of abundant, unpolluted water—a notion not easy to assail these days. But, like earlier pioneering Vermont decisions to ban billboards and threaten highway litterers with $500 fines, it has put a tangible premium on unspoiled nature, on the beauty of the land.

Vermont tradition is full of stories about Vermont farmers, including those who patiently, if somewhat dryly, give directions to leaf-lookers and other tourists from the city who gawk around on backcountry roads. These days they like to tell one about the New Yorker in the big shiny car who came up looking, somewhat belatedly, to speculate in land. After driving around for a while, he finally spotted a perfect, slightly run-down-looking farm hidden back in the hills. Finding the farmer out behind the barn at his chores, the visitor made a few friendly remarks, then with a sweep of his hand at the hillsides offered to buy $1,000 worth. "Ayuh," said the farmer, thoughtfully, as Vermont farmers are supposed to do, then looked up with a mildly triumphant smile: "Go fetch your wheelbarrow and I'll fill her up for you."

NATURE WALK / A Day at Lost Pond Bog

PHOTOGRAPHS BY ROBERT WALCH

Of all New England's natural features the most curious and compelling are its upland bogs. Unlike so many lowland swamps and marshes, which have been paved over or filled in with garbage dumps, these higher, remoter relics of the ice age have been largely bypassed and left to live out their natural lives. And they are fascinating lives, involving unique communities of plants and animals that thrive in mutual dependence, changing subtly but inexorably as the glacial ponds that gave them birth are slowly reclaimed by forest.

One of these ecological gems is Lost Pond Bog in the Green Mountains of southern Vermont, a place I visited one day in early June when it was in the full stirrings of spring. My guide was Hubert Vogelmann, a botanist and conservationist, whose own personal passion for bogs dates back to his graduate-student days.

To get to Lost Pond, we hiked from a forest road several miles up the Appalachian/Long Trail, then ascended a faintly marked side trail, passing through forests of sugar maple, beech and yellow birch that gradually gave way to red spruce and balsam fir. As we topped a rise at about 2,700 feet, the air became noticeably cooler and the ground damper underfoot. Through a tangle of underbrush and trees we caught the first glimpse of our goal: a forest opening of perhaps four acres, nestled below two low, rounded peaks and floored by a rich green carpet of brush, moss and sedges that curled around patches of open water dotted with lily pads.

One of the most intriguing aspects of a mountain bog, Hub Vogelmann explained, is that it exists in its own pocket of perpetually moist, cool, acid conditions. Thus it is an ideal meeting place for species that one would normally expect to find not in New England but much farther north —and even more surprisingly, for a few that probably originated much farther south. By all odds the most spectacular of these is the northeastern pitcher plant, one of three species at Lost Pond that eat meat to supplement their diets.

Picking our way out through the mucky forest surrounding the bog, we emerged on a springy mat of green sphagnum moss that circled the water, and almost immediately came upon a pitcher plant—a gaudily tropical-looking cluster of red-veined leaves, each 8 to 10 inches long and shaped like a tapering pitcher with a flaring cowl. This par-

THE APPROACH THROUGH THE BOG FOREST

agon of adaptation not only manufactures its food by photosynthesis like other plants, but makes up its need for nitrogen—of which there is always a shortage in the acid bog—by catching a little extra on the wing. Insects are attracted by the red lip of the pitcher, which secretes a sugary substance; if they venture in-

A PITCHER-PLANT LEAF

side, slimy walls and downward-pointing hairs make it difficult, if not impossible, to clamber out. The victims eventually drown in the pool of water the leaf has accumulated from rain; as each body sinks to the bottom, enzymes go to work to reduce it to a pulp. To demonstrate, Vogelmann slit open the tapered base of a leaf and exhibited the partially digested remains of a dozen insects. The species, *Sarracenia purpurea,* he explained, probably first evolved its

A RED EFT

A SUNDEW TRAPPING AN INSECT

A WOOD FROG

distinctive habits in the boggy, acid swamps of Florida or Georgia; from there it gradually worked its way north, its seeds most likely carried on the mud-caked feet of birds. While its relative, *Sarracenia flava*, reached its northern limit in Virginia, *S. purpurea* proved more resistant to cold, and dwells in bogs as far north as Labrador.

To locate the bog's second meat eater, another probable migrant from the South, we had to peer closely into the sphagnum mat, for the sundew is tiny in comparison to the pitcher plant, though no less ingenious in its lethal ways. It is well named: each leaf, scarcely half an inch across, is an innocent-looking little yellow sun bursting with reddish rays and tipped with a drop of innocent-looking dew. The "dew" is actually a sticky secretion that snags the legs of exploring insects. One obligingly fell into a sundew's grasp while we were watching. Slowly, over a period of several hours, the hairlike tentacles would arch inward to enfold the insect and extract the juices from its body, then unfold to be ready for the next snack.

Not all of the bog's denizens are so morbidly fascinating; a good many have a different kind of charm. While we were poking around in the thicket at the edge of the bog forest, we saw a little red eft, a shy, orange-red salamander that was scuttling to the safety of shade and drier ground. Not far from the eft we found another bog amphibian, a handsome wood frog with a black eye-stripe like a robber's mask, from which he

THE BOG MAT AND ISLANDS

gave us one curious glance before turning a profile and posing unblinkingly for photographs.

The Quaking Sphagnum

To keep our boots dry for the walk home, we had taken them off and left them on a fallen log. I soon discovered that going barefoot in a bog is the only way to appreciate its soft textures and squishy coolness, especially in the growing heat of a hazy June forenoon. As we walked across the shiny emerald rug of sphagnum, laced with the dense, wiry stems of leatherleaf bushes rooted in it, the water gurgled up cold and wet between our toes. It was indeed a "quaking" bog: one little jump was enough to set the supple mat in motion for 8 or 10 feet around. I dug into the green moss with my fingers, exposing a brown tangle of stems that marked the portions of the plants that had recently died away. The floating raft of sphagnum, which had slowly spread out from its original anchorage on shore, had probably built up to a thickness of several feet; as the weight of new growth on the top continued to push the raft down, and as the oldest pieces disintegrated and fell off beneath, the raft would finally become deep enough to meet the rising bottom, its weight gradually compressing the older debris into what would become pure, burnable peat.

It is this sphagnum, or peat, moss, Vogelmann noted, that is the key ingredient in most bogs. In a pond like this one that has no good natural outlet to keep flushing the water clean, the sphagnum begins to take hold, and once it does, its high content of organic acids drastically increases the acidity of the stagnant water and decaying muck around it. This sets the stage for acid-tolerant northern shrubs like leatherleaf and Labrador tea, and such trees as black spruce, larch and balsam fir. The acid also inhibits decay, so that dead matter piles up on the bottom and thus allows more plants to take root and fill up the pond. It is the sphagnum, moreover, that is largely responsible for creating the moist, cool conditions that also favor the northern species. The moss, whose cellular construction allows it both to float and to hold many times its weight in water, acts like a big sponge, continually absorbing moisture and losing it by evaporation, thus lowering the temperature of the water and air; the heavy cool air tends to get trapped over the bog, held in by the higher ground around it.

As we approached the open center of the bog, we could see that the sphagnum mat had invaded all but a half acre or so of water, which was broken into little bays where chunks of the mat had come loose to form islands. In places the moss had not yet reached, submerged plants like eelgrass and pondweed could be seen underwater, floating plants like lilies had rooted, and emergent plants like sedges had sprouted from shallower spots along the muddy bottom.

At the edge of the mat where it was firm enough to hold his weight, my companion reached out into the water and scooped up a gelatinous, iridescent blob, one of the many

A MASS OF FROG'S EGGS AND TADPOLES

A GREEN FROG

BLADDERWORT AND A WATER STRIDER

LEATHERLEAF, PINK BOG LAUREL AND SEDGES AT THE WATER'S EDGE

masses of frogs' eggs that drifted in the shallows; in this one most of the eggs had hatched but we could still see a couple of minuscule tadpoles struggling to break out. Nearby, one of the tadpoles' parents, a large green frog, set up a throaty *gunk, gunk* that sounded like an untuned banjo string; a spring peeper hidden in the brush tried out a few tentative notes as if warming up early for the evening chorus to come. As we walked on, the bog mat contributed some weird music of its own, exhaling pockets of trapped air beneath our feet in soft, crying protests, eerie whistles and low groans. In one little cove a pair of lozenge-shaped whirligig beetles chased and bumped each other on the water like Dodgem autos at a county fair; a long-legged water strider skated by to watch, dimpling the surface without slipping through. As the heat of midday mounted, so did other insects, mainly mosquitoes and black flies, and we slathered fresh bug repellent on our bare feet, hands, faces and necks with only partly successful results. Overhead, banks of cumulus clouds hung lifeless in the sky; it looked like a downpour in the making, but the day droned on, still and humid, and the deluge never came.

The Fearsome Bladderwort

Here and there in the water I had noticed what looked like fine strands of lacy seaweed floating partially submerged; I had scarcely given them a second thought when Hub Vogelmann announced, in a believe-it-or-not tone of voice, that I was looking at the bladderwort, the

THE BOG THICKET AND FOREST, SEEN FROM THE MAT

third and final member of the bog's terrible trio of meat-eating plants. His description of its eating habits was indeed the most unbelievable of them all, until I saw it for myself with the help of a pocket magnifying glass. The submerged, free-floating strands, only about an inch across and two to three feet long, were studded with rows of tiny transparent bladders, each perhaps a tenth of an inch in diameter. In its normal position the bladder is a flattened sac; when a water flea or other minute swimming organism touches one of the sensory hairs around its tiny mouth, however, the side walls of the sac instantly spring outward, sucking up water, organism and all through its trap-door valve. Under magnification, the victim can be seen swimming frantically around inside until it finally succumbs; meanwhile, the bladder slowly returns to its original flattened shape to prepare for another catch.

We continued our circuit of the bog and found, near the forest edge, an advance guard of taller shrubs —mountain holly, withe rod, purple chokeberry—forming an intermediate zone between the trees and the lower growth along the mat. In places an adventurous balsam fir had gotten ahead of the procession and marched out onto the acidic moss; it stood alone and forlorn, looking a little yellow and anemic from lack of nourishment, with a branch or two rusted brown from winterkill in its exposed location. Scattered about as we walked were the pink blossoms of bog laurel, one of the loveliest residents of the bog; nestled in shady spots near the thicket were wild callas, their broad white spathes clasped around stubby spikes, or spadices, crowded with tiny green dots that are the plant's real flowers. Above the bushes, brightly colored damselflies darted back and forth in search of smaller insects, settling momentarily on leaves or twigs and gracefully folding their wings; dragonflies, in contrast, rested between excursions with their double pairs of wings outstretched.

BOG-LAUREL BLOSSOMS

A DAMSELFLY

A DRAGONFLY

WILD CALLA

Bogged Down in the Forest

Having almost forgotten lunch in the course of our discoveries, we decided to cross over into the shade of the bog forest, stop to eat, then explore some more and eventually make our way back to the trail. My companion cautioned me that the wooded part of a bog is actually more treacherous than the tight, springy sphagnum mat we had been walking on, since conditions are less acid and there is more decomposition. As we passed through the thicket into the trees I promptly proved his point by slithering off a mossy log and quickly sinking in muck up to my knees. Helping to haul me out, Vogelmann cheerfully observed that we had almost lost me; to illustrate, he forced a long, dead branch a full six feet down into the gurgling ooze, a distance I calculated would have left me with only my eyebrows showing—if that much. While I was drying out on a hummock and eating my sandwich, Vogelmann remarked that, if it was any consolation, I would have wound up remarkably well preserved: thanks to the antiseptic qualities of the acids formed by accumulating peat, some well-known Danish bogs have

WILD LILY OF THE VALLEY

CLINTONIA

WILD SARSAPARILLA

CINNAMON FERNS ON THE MOSSY FOREST FLOOR

yielded bodies still fully clothed and only slightly brown and puckered after 3,000 years. In the old days, he observed, this was a convenient way to get rid of people: while ordinary thieves were simply hanged, those guilty of more heinous crimes were often disposed of in the local bog.

A Woodland Garden

I stepped with some care as we moved through the bog forest, which from the looks of it had already reduced the bog proper to perhaps half or a third the size of the original glacial pond. At the bog's outer edges, the ground was higher, with soil built up from innumerable generations of rotted vegetation and fallen trunks; here the living trees had also helped to dry out the ground by drawing up water and transpiring it to the air. In the cool forest shadows there were spinulose wood ferns, their fronds edged with thousands of spiny serrations, as well as cinnamon ferns, so called because the spore-bearing leaves, which appear in late spring, are densely covered with velvety brown hairs. Fossils of the latter, Vogelmann remarked with delight, proved that this type of fern had existed unchanged for over 250 million years. Strewn about the forest floor were small blossoms of all descriptions: the dainty white blooms of starflower, wild white violet and goldthread; the spraylike plumes of Canada mayflower or wild lily of the valley; and in higher, drier places back in the forest the greenish-yellow trumpets of clintonia, or bluebead lily, and the round white bursts of wild sarsaparilla. Everything seemed to be growing in harmony on everything else: lichens of various kinds clinging to tree branches, trunks and rotting logs; little seedlings of yellow birch and balsam fir taking root on old stumps or on the upended root balls of fallen trees. Overhead, from time to time, we could hear the piercing cry of a flicker, the melody of a yellowthroat, or the dull tacking noise of a red-breasted nuthatch picking bugs out of a tree trunk.

BIRCH AND FIR SEEDLINGS ON A STUMP

By now it was late afternoon, and we made our way out of the forest for a last look at the bog before meeting up with the trail and going home. The slanting sun cast a delicate, rosy-gold light on the surface of the water and its dark-green lily pads. For a moment we stopped and watched, and I thought about the whole remarkable community of life that surrounded this water, living on and above and beneath it, and how it was late afternoon, or evening, in the bog's own life as well. Happily, Lost Pond Bog is still lost to all but a handful who know what and where it is, and it is this feeling of unspoiled life and quietness that gives the place a timeless quality. What lends it particular poignancy, however, is that time is running out. As my companion noted, the process by which the forest claims its own starts slowly, imperceptibly, with the first tentative rootings of water plants. But it speeds up toward the end, as the growing bog plays host to trees. Lost Pond, already many thousands of years old, will become quite literally lost, swallowed by forest, in a mere century or two.

FOLIOSE LICHEN ON BLACK-SPRUCE TWIGS

LATE AFTERNOON AT THE BOG

3/ New England's Rooftop

One stately mountain there is, surmounting all the rest, about four score miles from the sea.

JOHN JOSSELYN/ *NEW ENGLAND RARITIES DISCOVERED*, 1672

It was a few minutes after five in the morning, on the top of Mount Washington, crown jewel of New Hampshire's White Mountains and the highest point of land in the northeastern United States. On this bleak summit, usually shrouded in mist and whipped by wind, a small miracle seemed to be taking place. The air was bright, clear, almost incredibly still. Over Maine to the east a red sun lifted slowly, then burst up through stubborn tatters of fog. Glowing orange, then molten yellow, it caressed the bald pates of Washington's neighbors, Jefferson, Adams and Madison to the north, searched out the tiny mirrors of a pair of alpine lakes to the south, and finally turned the whole gray, rocky landscape into gold. The handful of high peaks, like Arctic islands rising out of a temperate sea, were the only landfalls to be seen. For a hundred miles in all directions the lesser realms of New England lay somewhere down below, drowned in a brilliant ocean of white clouds.

Leaving the sun shining on New England's rooftop, I walked back through a snow patch to the Mount Washington Observatory, the weathered, gray shingle building on the summit where I had spent the night. My host John Howe, who has spent most of his professional life on the mountain since we went to school together years ago, already had coffee bubbling on the stove. "I thought this was the place that had the worst weather in the world," I said, recalling a well-publicized day in April 1934 when the breezes on Mount Washington averaged 129 miles per hour and one gust hit 231—the highest wind velocity ever recorded. "It is, but we have some nice days too," John replied with a smile, gathering up notations for his three-hourly report to the National Weather Service. "Heavy undercast, though; they must be having some drizzle in the valley." Very few people, I thought, except those who live on mountaintops, are privileged to use words like "undercast."

Even they, however, are not privileged to use it very often. Three quarters of the time Mount Washington, like New England's other highest peaks, has its head completely in the clouds, sampling the full flavor of some of the most variable and cussed weather in the United States. "If you don't like the weather," goes an old New England maxim, "wait a minute." On this mountain you cannot always afford to wait a minute; here weather is not so much a topic of idle conversation as a dominant, and occasionally fatal, fact of life. Along all of the many hiking trails that reach up into the Presidential Range there are stark metal signs erected by rangers of the White Mountain National Forest: *STOP. The area ahead has the worst weather in America. Many have died there from exposure, even in the summer. Turn back now if the weather is bad.* When the temperature in the valleys is crowding 85° and there doesn't seem to be a cloud in the sky, the temptation has proved strong to shrug at the signs and go for a walk in sneakers and a light shirt. This can be the ultimate mistake. More than 30 people have died on Mount Washington as a direct result of the weather, most of them from what scientists call hypothermia, the stupefying plunge in body and brain temperature that can result from a combination of wet clothing and chilling winds, even in moderate mountaintop temperatures of 40° or 50°—at 40°, for example, a 50-mile-per-hour wind sends the chill factor down to a numbing 10°. Knowledgeable hikers do not set out, even in August, without rain gear, sweater, wool cap and gloves stuffed into their knapsacks, and a flashlight for good measure. Those who don't are commonly referred to on Mount Washington as "goofers," and they have cost rescue teams more than one sleepless or tragic night.

To goofers, and even to experienced mountain men, it comes as a firsthand surprise that this ancient geologic bump, which in terms of altitude would barely qualify as a foothill in the Rockies or Alaska, does indeed, with the possible exception of some isolated peaks in the Arctic or Antarctic, have as severe a combination of wind, storms, cold and icing as any mountain in the world. The reason is that it lies high and exposed, with few surrounding mountains of any size to act as windbreaks, and it stands astride two of North America's most troublesome

The early light of a subzero winter sunrise spotlights Nelson Crag, a knob just below the summit of Mount Washington, which is seen at right trailing plumes of snow whipped up by winds gusting to 95 miles an hour.

storm tracks, which zero in on New England before heading out to sea. One track has its origins in the Arctic, sweeping in over the Great Lakes from the west and north, the direction of the prevailing winds. The other weather system, less frequent but often more violent, is a tropical pattern, including hurricanes that surge up from the south and east. In the almost constant pulling and hauling of cold and warm fronts, the ridgeline of the Presidential Range acts as a natural barrier, concentrating the winds from either direction and sometimes producing spectacular results. The moving ocean of air strikes the lower slopes, which push it upward, squeezing it between the mountains and the masses of air above. The constriction produces a venturi effect—an increase in wind speed by a third or more as the air races to get through the narrowed opening above the summits. As the moist lower layers are forced up into cooler air, they reach their dew point and condense, forming clouds and often bringing rain, sleet or snow; the last has fallen in every month of the year.

The summit of Mount Washington, in effect, is a chunk of Labrador set down less than three hours' drive from Boston. As such, it has proved a magnet for scientists and sightseers alike. It has served as a unique instrument platform and a laboratory for testing everything from cold-weather clothing to the control of icing on aircraft and jet engines; in addition to these projects, and basic weather reporting, the observatory has contracts for measuring such exotica as the nature of fog droplets in clouds and the behavior of cosmic rays.

After breakfast we were joined at the observatory by John Nutter, director of education for the Appalachian Mountain Club, which has its north-country headquarters directly below the mountain in Pinkham Notch. John was to hike with me over the crescent-shaped line of peaks to Mount Madison six or seven miles north, along the highest section of the Appalachian Trail and through the largest area of alpine tundra in the East. By the time we shouldered our packs and said goodbye to John Howe, the weather was back to something approaching normal: the undercast of the early sunrise hours had lapped up the summits to become overcast—"incast" might be a better term. We set out in an eerie moonscape of gray-green boulders that loomed out of a soft, soaking fog, a Wagnerian stage set made to order for the ride of the Valkyries and lacking only Lauritz Melchior in full throat on a nearby crag. Hopping from rock to rock to avoid the deepest snow that lingered on the trail, we came to the edge of the Great Gulf, a huge bowl that plunges away from the crest of the northern Presidentials with all the authority of a first-class abyss. The gulf, John pointed out, is the largest of the series of amphitheaters formed along the eastern rim of the range during the last ice age, when snow blown over the summits by the prevailing winds *(picture, pages 74-75)* accumulated in the leeward ravines and gradually packed down into glaciers that gouged and plucked the V-shaped valleys into U-shaped bowls. The 5,552 acres of the Great Gulf, while tiny by Western standards, constitute the first official wilderness area in the mountains of the Northeast protected by federal law against timber-cutting, building and roads.

As we skirted the gulf, John jumped from boulder to boulder like a mountain goat, stopping every few moments to crouch and study a new botanical find. "A lot of people assume that nothing grows on this rock pile," he exclaimed in utter disbelief. "Do you know there are well over a hundred species of plants living up here—including some that grow nowhere else in the world?" I peered through the ribbons of fog. The landscape of sharp boulders, shattered out of the tough mica-schist bedrock of the summits by alternate freezing and thawing in the harsh postglacial climate, was not the wasteland it appeared at a distance, but a rich mosaic of tiny alpine plants. The stony surfaces of the rocks themselves were peppered-and-salted with lichens—bright green, dark green, gray, silver, black—their colors almost luminous in the saturating mist. Tucked in crevices out of the blast of the wind were little tufts of green club moss; strewn about in soft cushions were the red-green leaves of diapensia, which stand full exposure to the cruelest winds and every June still produce little bouquets of white flowers, each hardly bigger than a pencil eraser. Here and there were other plants whose names revealed their Arctic origin: Lapland rosebay, Greenland sandwort, Labrador tea. Among the smallest was alpine azalea, forming miniature mats less than half an inch high. Where enough soil had accumulated in the rock fields from centuries of slow weathering, there were grassy, waving stretches of sedge in large, level areas, referred to on Mount Washington as "lawns."

Because of the severity of wind and weather, John noted, treeline ends on the Presidentials at half the height it does in the Rockies, about 4,800 feet on the exposed western and northern sides; a little higher, 5,200 feet, in the shelter of lee slopes where the wind slacks momentarily to deposit a protective blanket of snow. On the lower shoulders of the summits we could see the advance guard of the forest that had tried, and failed, to reach the peaks: subalpine species like black spruce, balsam fir and paper birch, which normally grow 40 to 70 feet tall in the valleys, were camped here in crouching communities of *Krummholz* (literally, crooked wood), sculptured into stunted, windblown mats that ranged from three or four feet high in sheltered spots to a mere foot or less in the open. In places a brash young shoot or two had ventured above the mat; it would not be long before such pioneers were pruned back to size by the icy, desiccating winds. Up here, John observed, plants have to stick together to break the wind and hold the insulating snow; individualists rise above the crowd at their own risk.

Alpine flowers found in the White Mountains, as tiny and delicate-looking as they are tough, include moss plant, whose bell-like blooms are less than a half inch in diameter, and alpine azalea, which has quarter-inch blossoms. Mountain avens, a comparative giant, has one-inch blooms. Alpine bluet, a white variety of a meadow flower, exists only in the White Mountains and two islands off Newfoundland. The last is a true rarity: dwarf cinquefoil, so small that a whole plant is the size of a half-dollar, grows only on Mount Washington and the nearby Franconia Range.

The adaptations by which these plants are able to survive are remarkable, and have probably been studied more closely, by more botanists, in New Hampshire's Presidential Range than in any other place. Most of the species grow close to the ground, where they not only avoid the highest winds but also take advantage of the radiant energy of the sun, which heats up the ground surface and adjacent air: when the wind at head height is 30 miles per hour and the air temperature 45°, for example, in the ground-hugging alpine gardens the breeze may be only 5 to 10 miles per hour, the air temperature around the plants may be 50° and the soil itself an almost tropical 70°. Many of the plants have the leaves closely massed to deflect wind from their innermost parts; the leaves themselves are small and round or needle-shaped, presenting a minimum of surface area; they have waxy skins as further protection against the drying effect of the winds. Some plants, like Labrador tea, have woolly brown hairs on the undersides of their leaves, which serve to inhibit moisture loss and to collect fog droplets; Labrador tea also has a distinctive habit of curling its leaf edges down to provide still more shelter for the porous undersurfaces.

In other responses to the climate, such subalpine species as spruce retreat partway underground, developing root systems that are far more extensive than their exposed upper parts. Plants like mountain avens and Lapland rosebay, which depend on insects for pollination, produce masses of bright flowers to lure bees and other insects during their short flowering seasons in spring and summer. Spruce and fir, which depend on long-maturing, windblown seeds to reproduce, abandon the practice on high slopes because they simply cannot grow fast enough or store enough food to produce cones regularly. The higher balsam fir ventures, for example, the fewer cones it bears; at about 5,500 feet, its maximum elevation, it bears none at all, relying solely on the reproductive process known as layering, in which lower branches in the

MOUNTAIN AVENS

ALPINE BLUET

DWARF CINQUEFOIL

ALPINE AZALEA

MOSS PLANT

stunted mat droop to the ground, become covered with humus or moss at these points, sprout new roots and sometimes create a whole ring of little new trees around the parent. The success of all these adaptations can be measured in terms not only of tenacity but also of ripe old age; counts of annual tree rings in gnarled trunks frequently range from 100 to 150, and the root systems are often older. One botanist reported a dwarfed spruce tree still going strong after more than 360 years.

As we continued to explore Mount Washington's alpine gardens, the fog drip became heavier and the wind began to rise, so we put on our rain gear and settled in the lee of a craggy knob for lunch. Somewhere from the mist came the trill of a slate-colored junco, commonest of the few birds that make themselves at home in this Arctic zone. A little later we heard the croaking call of a raven, that oversized cousin of the crow, which after a long absence has returned to New England's highest peaks; the crew at the observatory delights in watching these big birds practice their soaring and hovering in the high winds of the summit, and have taken to leaving out scraps of food and photographing the birds as they sweep in to scavenge them. Aside from these two bird species, wildlife at this altitude is rather limited, although we did note tracks of snowshoe hare. In summer woodchucks, deer and even beavers sometimes wander up onto the heights for a look around.

The only permanent mammal residents at or above treeline are small, like the plant species—mice, moles, voles and shrews. Of these the smallest—in fact the smallest of all North American mammals—are the shrews. Fast-moving, sharp-nosed, beady-eyed insectivores, they live mostly in cool, moist crevices and patches of low shrubs or moss; because they are so tiny and quick, anyone who would observe them needs good eyes and a great deal of luck. Of the six species found on and around Mount Washington, the long-tailed shrew, about five inches long including a two-and-one-half-inch tail, is the one most likely to be discovered near the summits; a cousin, the short-tailed shrew, is distinguished by having a poisonous saliva that can immobilize such larger opponents as mice, which it does not hesitate to attack. A third, the pygmy shrew, is the smallest of all: a full-grown adult is almost entirely fur and so fine-boned it weighs little more than a dime. The water shrew, which inhabits the borders of lower lakes and streams, is perhaps the most fascinating: it can actually walk on water, helped by its light weight, speed and fringes of stiff hair on its feet that allow it to scamper across the surface without sinking. All shrews have one trait in common, an astronomic rate of metabolism that can exceed 1,200 heartbeats a minute. They dash about constantly, twittering, in search of anything to satisfy their voracious appetites—if not insects, then mice, a probing finger, even other shrews. A shrew is definitely not for taming: about the only other creature it can get along with is its mate, and then only briefly during breeding season, which in this fast-living animal occurs two or three times a year; if captured and confined, a shrew becomes even more frenetic than usual and may literally worry itself to death in a matter of minutes from a sort of shock. Even if the shrews escape their natural predators, such as weasels or hawks, they lead hurried, furious and scarcely enviable lives, burning out at the extreme old age of 14 to 16 months.

Musing on these odd, omnivorous midgets of the mountains, John and I packed up the remains of our lunch and returned to our walk. As we headed over the shoulder of Mount Jefferson, John suddenly decided to release some excess energy of his own by demonstrating skiless skiing, sliding on his boot soles down a snowfield that dropped away sharply into the fog in the general direction of the gulf. "Glissade," John explained, giving the technique a proper French flourish. "Lovely," I replied, in English, but I preferred to take pictures of some lichens on a securely anchored piece of quartz. Trotting back up with scarcely an extra breath, John recounted the story of a girl who had broken an ankle glissading not too far from here; it had taken a rescue team, John included, several hours to get her out on a litter, three men on each side. To spell the litter bearers, the team conscripted 40 Green Berets who happened to be on the mountain for an early season training exercise. "They were in such lousy shape," John confided, "that they couldn't do much carrying the last mile or so."

Beneath Jefferson, in a low saddle called Edmands Col, we passed a squat Quonset-type shelter of corrugated steel, placed halfway between Washington and Madison as an emergency refuge. This reminded John of another story; he recalled that one member of a boys' camping group caught in a sudden storm had begun to slip into the shivering, mumbling daze of hypothermia. Fortunately, he said, a young woman medical student taking cover in the shelter recognized the signs in time. She stripped the boy, ordered another boy to strip, and hastily piled them together in a sleeping bag until skin-to-skin warmth brought the near-victim around.

On the ascent to Mount Adams, there were other clues to the severity of weather in the Presidentials. John called my attention to the cairns piled every 50 feet or so to mark the trail—an indication of how bad the visibility can get and how thoroughly lost hikers can become in these rock fields when caught in blinding rain, snow or fog. Beyond "Thunderstorm Junction," a meeting of several side trails marked by an oversized cairn, some considerate early trail builders had fitted flat rocks together to fashion a level path across a side slope, saving goofers some nasty slips and falls. As we clambered down through a patch of spruce growing four or five feet high in the shelter of a hill, we were startled by the high sweet melody of a white-throated sparrow, one of the few other inhabitants of the Alpine zone; his song ("O-o-old Sam Peabody, Peabody, Peabody" or "O-o-oh Sweet Canada, Canada, Canada," depending on where you come from) is one of the finest in New England and distinctive enough even for a novice to recognize. We spotted the bird on a branch near the trail, puffed up with his bill pressed down in his chest feathers against the cold, blowing fog. Almost directly beneath him there was an opening tunneled back among the stunted firs, where broken branches and a brown, scarred area of compacted earth indicated that successive groups of hikers had found a cozy campsite and had begun to alter the landscape bit by bit. The Forest Service, John observed, had finally had to prohibit all camping and fires above the main treeline for this reason, before the fragile arctic vegetation became trampled beyond its ability to recover. "It's too bad," he said. "There are just so many people coming up here to enjoy the mountains these days. We ask them not to cut boughs for bedding or firewood, not to travel in large groups, not to wash at drinking-water sources, not to pick the flowers, not to leave litter that others will just have to pick up and carry out. We've only been partly successful so far. With all the people using Tuckerman Ravine, one of the most popular spots for late-spring skiing and hiking, last year it cost ten thousand dollars just to clean up the place and haul the accumulated garbage out."

After six hours exploring the crestline of the Presidentials, we arrived at the Appalachian Mountain Club's Madison Spring Hut, where I was to learn a few more facts about modern mountain life. The "hut," a sizable and sturdy building securely anchored against the winds, is equipped with a big kitchen, wooden tables and benches and two dormitory rooms that can sleep 50 people in four-decker bunks; it is not something that one might ordinarily expect to find on an alpine mountaintop, but many a cold and hungry hiker, I was sure, had been glad that it was there. Originally built as a small stone cabin in 1888, it is the oldest of the A.M.C.'s chain of nine mountain hostelries, strung out

A masked shrew, little more than four inches long from nosetip to tail end, feasts on a grasshopper. Six species of these voracious little mammals live on or near Mount Washington, mostly in crevices under rocks and plants.

through the White Mountains a day's hike apart. Together they provide hot food, bedding and a chance to travel light for some 35,000 hikers a year, plus stopping places and emergency shelter for countless others. The huts—the largest, at Lakes of the Clouds just south of Mount Washington, can sleep 90 people—have come a long way since a band of Boston professors and outdoorsmen founded the A.M.C. in 1876; some of the club's 20,000 members, in fact, are beginning to wonder how they got into the "hotel business" anyway. Over the years, however, the huts have become so popular that several are now opened earlier and closed later than the normal season, and two, including the club's main base camp at Pinkham Notch, are used by climbers and hikers throughout the year. Along with the A.M.C.'s admirable battery of regional maps and guidebooks, and an educational program that includes everything from mushroom-hunting to first aid, they have helped uncounted people to gain a close acquaintance with New England's mountains.

By the time we arrived at Madison that evening, a dozen hikers had already converged on the hut, including kids from Pittsburgh and Boston, two health-clinic workers from Oswego, New York, a research librarian from Buffalo and a couple from Canton, Connecticut, who had carefully packed along a small birthday cake and candles for their daughter Jennifer, aged 10.

After supper I had a chance to add more details to my growing collection of mountain lore. Mount Washington, I was beginning to realize, is not just a mountain, but a phenomenon, an institution and something of a paradox as well. It is certainly the granddaddy of American wilderness areas; people were rock-climbing and looking at flowers here long before the West was won. It can also lay claim to being the most climbed, most studied, most lived-on mountain in the United States, if not the world. The first recorded ascent dates back more than 300 years, a good two centuries before Switzerland's Matterhorn was first climbed in 1865. In June 1642, Darby Field of Exeter, New Hampshire, lured by the glittering summits of what were then called the Chrystall Hills, rode into the trackless interior and persuaded a couple of reluctant Indian guides to accompany him to the top of their sacred mountain, Agiochook. The story has it that Field was in search of precious stones, though the sparkling "crystals" he saw were undoubtedly nothing more than the sun shining on the millions of bits of quartz and mica imbedded in the summit rocks; to his dying day he is said to have ranted about the treasure he knew was there but never found.

By 1784 the first scientific party to reach the summit officially changed its name to Washington; in 1819 Abel and Ethan Allen Crawford, enterprising early innkeepers in the valley, blazed the first trail to the top. A year later some of the leading citizens of Lancaster, New Hampshire, achieved the summit with a supply of something called "O-be-joyful" and proceeded to christen the nearby peaks, in order of height, after succeeding Presidents, toast by toast. By the mid-1800s, before scarcely anyone had even heard of Yosemite or Yellowstone, the White Mountains had entered their golden age as the "Switzerland of America" and the valleys were dotted with an increasing array of elegant Victorian hotels. An eight-mile carriage road was completed up one side of Washington in 1861, a cog railway up the other side in 1869; one of the early passengers on the latter was President Ulysses S. Grant. The circus impresario P. T. Barnum, a frequent visitor to Mount Washington, generously pronounced the view the "second-greatest show on earth." The opening of the carriage road in particular set off a craze for ascending, or descending, the mountain in every way the human mind could devise: walking, running, barefoot; on skis, snowshoes, dog sleds, even tricycles; and later on motorcycles and in all manner of vehicles from Stanley Steamers to high-powered racing cars.

Meanwhile, lumbermen were busy too, building railroads of their own into the innermost valleys and systematically stripping the region of trees. Appalled by the growing sight of scalped scenery, and by several disastrous forest fires, such organizations as the A.M.C. and the Society for the Protection of New Hampshire Forests, along with Governor Frank Rollins and U.S. Representative John Weeks, fought for passage of the Weeks Act in 1911, which made possible the purchase of watersheds and timberlands for the national forests in the East. Today the Victorian hotels are mostly burned down or boarded up, replaced by motels and Winnebago campers, backpacks and freeze-dried foods. The White Mountain National Forest, grown green again, embraces 729,000 acres (more than 80 per cent of the land within the ultimate boundaries originally outlined for purchase, including 46 peaks over 4,000 feet and some 1,200 miles of hiking trails; this contrasts with a total of 245,581 acres in the Green Mountain National Forest in Vermont, where government land-buying started later and has still accounted for only 40 per cent of the area within its purchase boundary). The White Mountain National Forest comprises a land area larger than that of Rhode Island and, as the largest single piece of public land in New England, one

The crisp lines of cube-shaped granite blocks

lying near New Hampshire's Crawford Notch indicate that they have only recently split off and fallen down from the cliffsides above.

almost as intensively used; it is within a day's drive of more than 50 million people, a quarter of the population of the United States. A good 200,000 tourists get to the top of Mount Washington each year, by the auto road, cog railway or on foot. Forest rangers, worried lest the Appalachian Trail become the Appalachian Trench, are "hardening" easily worn-down sections with a variety of techniques to minimize the destructive patter of lug-soled boots. To spread the recreation load, attempts are being made to open up lesser-known backcountry areas, and to give them wilderness protection against logging, roads and vehicles of any kind. This has not pleased certain paper companies in need of wood, nor has it appealed to old-line outdoorsmen, who are sure a "wilderness" banner flown anywhere will only bring more eager hordes of campers in to destroy everything in their path.

That night I tried to grapple with the growing management problems of wilderness, and lost. When I finally got to sleep, I dreamed instead that I was Toni Matt, the great Austrian skier, who in 1939 schussed the headwall of Tuckerman Ravine without a single turn, while thousands cheered below.

The next morning we left Madison Spring Hut, still enveloped in a gray-green mist, passed by a little alpine puddle named Star Lake, then abruptly crossed a brow of rock called the Parapet and went down into the gulf. The path we followed, the Madison Gulf Trail, is among the more direct in the White Mountains—meaning vertical. It drops over and around house-sized boulders and down a headwall hollowed out by one of the tributary glaciers that joined to form the Great Gulf. The firs and spruces, now in the lee of the mountain, began to stand on their feet again, twining their roots around rocks and into crevices, extending sinuous, slippery tentacles across the path. The rock faces, too, were slick, from moss and melting snow; on the more challenging parts I used hands, feet and seat of the pants, while John was not looking, to put on the brakes. At the foot of the upper headwall we negotiated a chute called Mossy Slide, then entered a dark forest glade where the stream we were following, Parapet Brook, slid gracefully over a smooth rock wall in a glittering veil. Usnea, old-man's-beard lichen, clung in shaggy patches to the trees; big boulders by the stream wore shining wigs of moss on which whole settlements of saplings had taken root. While I rested gratefully and adjusted an aching pack, John burned off some more excess energy by practicing rock-climbing up a sheer precipice of stone, hanging happily by his fingernails.

On the trail again, we pushed through a maze of fallen trunks, crossed and recrossed the brook and descended the lower headwall into thick woods. Here we began to see goldthread in flower, carpets of wood sorrel, the fiddleheads of uncurling ferns. Here too were painted trilliums, among the most beautiful of woodland flowers—each with three white, delicately tapered, outstretched petals, veined purple at the throat, above three echoing dark-green leaves, offset to form a six-pointed design. Everywhere the forest was bright green with the viburnum called hobblebush, its bold flower clusters bouquets of showy sterile outer blossoms around masses of small fertile blooms. We heard another white-throated sparrow, followed by a medley of warblers and a single winter wren. Then, coming down over a low rise, we abruptly left the mountain mists behind. To the south, stretching up to the Presidential crest where we had been, loomed the whole deep enclave of the gulf, its headwalls and summits now far above us, darkly stenciled against the afternoon sun.

When we finally made our way back to the A.M.C.'s Pinkham Notch Camp in the valley, we unlimbered our packs in a veritable locker room of activity in the front yard. Bare-chested rock and ice climbers were tromping in off the mountain, brandishing ice axes, crampons and coils of nylon rope. Late arrivals from the city, their gunfighter mustaches twitching in anticipation, were lashing boots and skis on top of bulging backpacks, heading for Tuckerman to camp overnight and ski the headwall the next day. On the porch a bearded hearty with a zither and some time on his hands was serenading a couple of long-haired blondes. I barely had time to wash the rock dust out of my teeth when the large iron gong in the yard sounded line-up for one of the A.M.C.'s staggeringly healthy meals.

Later, before turning in, I could see the moon touch the mountain briefly through the clouds. Above the soft cascading of the Cutler River back of camp rose a low, steady roar. It was the ocean of air, still assaulting the summit from the west, breaking up over the rocks in an endless wave and washing down through the branches of a million trees. For all the traffic, Mount Washington was still wild.

Such Lovely, Savage Hills

PHOTOGRAPHS BY DEAN BROWN

From 20 miles out in the Atlantic they loom, on a clear day, like soft white clouds. There is little of the jagged drama of the Alps or Rockies about the White Mountains of New Hampshire: they have been worn down through the ages to perhaps half their original height. Even the tallest peaks today, the Presidential Range, are not much more than a mile above sea level. Yet for all their rounded, apparently gentle aspects, the White Mountains, when experienced close up, present some of the ruggedest alpine scenery and conditions in the world.

The dominant forces in the shaping of these mountains are generated by climate. Standing at a point where masses of cold, dry, subpolar air from the north and west clash with moist, warmer air from the south and east, the mountains are subject to heavy rains and snows, as well as sudden temperature reversals that result in hard freezes and abrupt thaws. The highest summits experience some of the coldest temperatures and most powerful winds on earth: on top of Mount Washington, the thermometer routinely plunges below zero and gusts of over 100 miles per hour have been recorded in every month of the year.

Climate has determined virtually everything about the White Mountains. The weathered, sparkling mica schist that, together with a long-lasting snow cover, gave these peaks their name has been blasted for millions of years by wind, rain and snow, shattered by frost and brought down in landslides. The same climatic conditions determine the kinds of plants that can cling to these peaks —a subalpine vegetation made up of stunted firs and spruces, and low alpine shrubs such as Lapland rosebay usually found only at much higher altitudes or much farther north. Finally there is water everywhere, pooled in peaceful highland lakes or rushing in fierce torrents down the valleys after spring thaws and drenching summer rains.

It is these elements that have made the White Mountains a spectacularly beautiful region—and that can make them a dangerous place. Dean Brown, who took the pictures on these pages, was an experienced outdoorsman. But, climbing Table Mountain toward the end of his assignment he was tragically betrayed by slippery, frost-cracked rock and fell 100 feet to his death. His last picture, taken of that rock toward sunset, is on pages 98-99.

Mist-hung, cloud-shrouded, the valley of the Zealand River southwest of the Presidential Range evokes the elemental forces that pervade the White Mountains: variable weather and rushing water in a setting of tumbled rock and dark evergreens.

Mosses, lichens and firs share living space among bedrock outcroppings in the Great Gulf.

The Difficult Life of Mountain Trees

In their struggle for existence on steep, rocky slopes, the forest trees of the White Mountains achieve strange feats of growth. Along the Wamsutta Trail in the Great Gulf *(left)*—a huge trough excavated by a glacier some 12,000 years ago—fir and spruce may grow to 70 feet in height, although their roots may find only precarious toe holds among the jumbled rocks. Aiding the trees to get started are the lichens and mosses that grow abundantly in the moist ravines: the lichens help to break down the rock faces, releasing nutrients to the forest floor, and the mosses act as nursery beds, incubating the seeds dropped by the trees until the saplings' roots have grown long and strong enough to wrap around the rocks and push down to find support and nourishment in pockets of soil below.

Above treeline at about 5,000 feet, the struggle for existence is even more dramatic. Here, out of the shelter of rock faces and buffeted by frequent gales, the same species found farther down have shrunk to form dense mats of twisted, stunted trees called *Krummholz;* on Mount Washington wizened black spruces may be a half century old yet only one foot tall. Within the protective windbreak of this mat, close to the ground, grow small shrubs, sedges, lichens and mosses similar to the vegetation of arctic tundra beyond the continental treeline some 1,000 miles farther north.

In a dense mountaintop mat of Krummholz, black spruces dwarfed by the harsh climate are no taller than foot-high boulders among them.

At 2,500 feet in Zealand Notch, big paper birches thrive in a highland forest dotted with mountain aster, bunchberry and wood ferns. So-called erratics, boulders carried by glaciers from sites farther north, litter the ground and make footing difficult even in this relatively placid setting.

Boulders, inched back each winter by expanding ice, line the shores of Spaulding Lake beneath a low-hanging canopy of mist and rain.

Levelers of Mountains: Water and Ice

The climate produced by the clash of weather systems above the White Mountains makes them a watery realm indeed. In glacier-cut cirques and valleys, rain water and snow-melt pour off steep, rocky walls to fill mountain tarns such as Spaulding Lake *(left)* in the bowl of the Great Gulf. Some of these lakes are shallow, some surprisingly deep, but all show, beneath their frigid, crystal surfaces and along their shores, the rocky litter of erosion from above. And the process continues: the walls of the Great Gulf and other cirques are still under constant attack by frost, wind and driving rain.

Ice, with its splitting, cracking action, is one of the great destroyers of these mountainsides, and it helps to shape the lakes below as well. In winter, when a lake's surface is frozen, the ice expands and contracts with temperature changes, pushing rocks in the shallows up onto the shore, sometimes building ramparts as high as 10 feet along the banks. (Geologists estimate that on a New Hampshire lake one 75-ton boulder has been moved 155 feet, in a zigzag pattern, over the past century and a half.) Thus each spring, when the ice breaks up, the contours of the lake will be a little different from the year before. And then as the spring freshets come, or the heavy rains of summer begin, the lakes fill up and overflow in a rampage that can flood and even alter the channels of their outlet streams *(overleaf)*.

Through ice-cleaved chunks of granite, the Zealand River wends its way in tiered cascades.

Running wild after unusually abundant June rains, the Ellis River in Pinkham Notch surges far beyond its banks, engulfing streamside

boulders and trees. When such violent floods occur, a stream may change its course, cutting new channels in its haste to get downhill.

The Ever-Changing Forms of Rock

To the casual hiker through the White Mountains, rock is the most stable feature of the landscape, seemingly unchanged and unchanging. That is an illusion, however, in a region where incalculable forces have long been at work. Some 360 million years ago, tremendous pressures folded the area's original sea-sediments like bread dough, and raised them up as tough schists. Later, molten granite welled up from below. Though violent change seems ended, these rocks are still vulnerable. The weathered slope at right, for example, is disintegrating slowly but constantly as a result of a process called exfoliation: rain water and snowmelt seep into cracks, then freeze and expand, breaking slabs loose from the surface.

Still more dramatic are the landslides that occur on steeper slopes: when the workings of frost and water have cracked and saturated enough rocks, a whole mountainside may suddenly lose its equilibrium and roar down in a deadly cascade.

The most striking example of the effect of climate, however, came at the closing of the ice age, when the scoured and barren mountaintops were exposed to fiercer conditions than now. Frost action literally broke apart the summits into piles of giant fragments *(pages 102-103)*; some of these rocks are creeping down into the valleys as "rock glaciers," slowly reducing the heights of the peaks still more.

Exposed, weathered granite faces glint amid woodlands on Table Mountain south of

the Presidential Range. Frost loosens the fractured surface until eventually it peels off in sheets, making for slippery, dangerous footing.

Near the summit of Mount Washington, layers of folded rock spotted with yellow lichens show how pressure and heat transformed horizontal beds of shale into contorted schist. On a much grander scale, the mountain itself is just such a warp in the earth's crust.

Debris of an old landslide in the Great Gulf is partly swallowed by forest growth. Landslides occur in the White Mountains when rocks, weighted with water and loosened by frost, break away over a large, steep area and then tumble destructively downhill.

A surrealistic landscape of bare rocks called Felsenmeer—literally, a sea of rocks—covers Mount Washington and distant peaks of the

Presidential Range. Heaved out of bedrock by frost, the huge boulders attest to the ferocity of the climate that followed the last ice age.

4/Varmints, Loons and Whiskey Jacks

Sound has no power to express a profounder emotion of utter loneliness than the loon's cry....You can fancy no response to this signal of solitude disturbed.

THEODORE WINTHROP/ *LIFE IN THE OPEN AIR*.1836

Everybody knows about the loon; he is a crazy bird that hangs around northern lakes making weird and foolish noises by the light of the moon. This is true, except for the foolish part; he is really quite an intelligent bird. He is also New England's oldest feathered resident, and, if you get to know him, one of the region's more fascinating creatures on many more counts than his voice alone.

To start with the voice, however. It is a sound without any precise equal in nature—one doubly treasured in New England, where the loon was heavily hunted and driven north for a number of years but now seems to be making a precarious comeback. The sound of the loon has been described by generations of delighted visitors to the north woods: mournful, mirthful, maniacal, unspeakably lonely, the embodiment of wildness, the soul of solitude. To the early Abnaki Indians it meant the messenger of Glooskap, one of their principal deities, who had given the bird a doglike voice; whenever a loon cried at night on a lonely lake it was to deliver a secret message to his master in the sky.

To later ornithologists, the loon's varied utterings were a challenge in classification. One of them, William Lyman Underwood, pointed out that the loon had not one or two but at least four distinct calls: a short, cooing note, often heard when several loons were together; a long, drawn-out *hoo*-ing, the weird and lonely night call; a rollicking yodel —the famous "laugh"; and the little-heard summons that old Indian guides referred to as the storm call because it was a sure sign of approaching bad weather. Underwood, an accomplished imitator of the calls of various birds, apparently trained his voice so he could coo and yodel with the best of loons; for the greatest accuracy, however, he reported that the ideal instrument was an old-fashioned ocarina, or sweet potato, key D, size 5½.

I bow to Mr. Underwood. I could not begin to reproduce the loon's wild, spine-shivering medley of sounds, even if I could find an ocarina. I did briefly try some imitations one summer night on a northern New England lake. I had just finished an evening swim and was restoring my circulation by the fire when the incredible performance began. First, the long, ghostly *hoo-oo-oo*. Silence. Then the yodeling, *ha-ha-loo,* etc. Silence again. Had both sounds come from the same bird? I stumbled through the undergrowth to the water's edge; there, without doubt, was the source of all the noise, bobbing happily on the moonlit surface like a big fat bathtub toy. Slowly he drifted out of sight around a little island, and from time to time the resonant calls came back. I whistled, hummed, yodeled; nothing came close. So I opened a can of pork and beans, settled down by the fire and listened into the night until the wonderful concert was done.

Someday I would like to spend a whole summer loon-watching, as well as listening, for *Gavia immer,* the common loon, is an uncommon and endlessly absorbing bird. Not the least intriguing fact about him is that he has been around longer than most birds, and much longer than people. Along with the similar-looking but unrelated grebe, I discovered, he has plied the northern waters virtually unchanged since the Eocene period, 60 million years ago; because of this impressive seniority he is accorded the place of honor in the front of most books on birds.

To last that long, one would suspect, the loon must have evolved some exceptional equipment, and he has. Beneath his gorgeously snappy summer plumage—black-and-white checkered coat and striped collar and an iridescent greenish-purplish head—he is a big, muscular bird the size of a goose, a powerful machine almost perfectly adapted to water. He certainly lives up to one of his other common names, great northern diver. To catch fish, his main food, as well as an occasional frog or crayfish, he plunges straight down from the surface like a torpedo, wings folded back, propelling himself with rapid thrusts from his paddle-like feet. If a burst of speed is called for, he opens his strong but relatively stubby wings and quite literally flies underwater, grasping the fastest of fish with his long, dagger-like bill. He often stays sub-

A common loon on a New Hampshire lake demonstrates its submersibility by sinking partway underwater, while its mate prepares to take off.

merged for a minute at a time and can go as long as three minutes in pursuit of deep-swimming or elusive prey; some birds have been accidentally snared in fishermen's nets at sea as far down as 180 feet.

Deep diving and powerful swimming are not the loon's only submarine skills. When disturbed or threatened by man, he can sink out of sight completely, leave only his head and neck exposed or swim along with his bill protruding as a breathing periscope. This ability stems from the fact that the loon is naturally heavy, with unusually solid bones that have few buoyant air spaces in them. When resting on the water he often appears just barely afloat; like a submarine he can rapidly change his buoyancy, blowing ballast by expelling air from his lungs and even pressing out bubbles trapped in his feathers to sink to the desired depth. If necessary, he adds further evasive tactics: when a family of loons is pursued, one parent will swim off in a different direction, uttering cries of warning, flapping, diving and reappearing to draw attention to himself. Sometimes he will double back on a pursuer, diving under a boat and popping up disconcertingly on the other side.

If the loon is a strong and graceful athlete on water, he is a bumbling oaf on land. His short legs, set far back for swimming, can barely hold him upright, and when surprised ashore he must lurch and stumble toward the water, using his wings and even his beak as props in a headlong scramble for safety. When he does reach his element, he seems to have considerable trouble getting out of it again; he has to flap along the surface, feet going like a treadmill, for as much as a quarter of a mile before his short wings can get his heavy body airborne.

Once aloft, a loon moves with the fastest of birds, often reaching speeds of 60 miles per hour. You can usually tell a loon from other goose-sized objects in the sky by the way he appears to fly hunchbacked, drooping at both ends, his downward-curving neck and head pointed rakishly forward and his feet retracted behind. When he is alighting, his speed and weight catch up with him and he comes in like a heavy bomber, crash-landing and sending up a line of spray across the lake until his momentum is slowed.

The loons first land on the northern lakes in early spring as the ice goes out, flying in from the nearby Atlantic, where they have spent the winter fishing and riding out the storms. The mated pairs, like other summer residents, seem hardly able to wait to get back to their favorite vacation spots (they like their privacy, moreover, and will often attack and chase off ducks or coots that have had the temerity to stake out part of their pond). The courtship dance that soon ensues is a sight to see: the male rushes about, running almost upright on the water with wings folded and bill open, then sinking down for a moment's rest before starting over again. The pair builds a nest that is primitive in its simplicity: a shallow depression near the lakeshore, on an island or atop an abandoned muskrat house, matted perhaps with a few soggy mosses and reeds. By late spring or early summer there are usually two olive-brown, dark-spotted eggs, and both parents take turns sitting on them until they hatch a month later. The downy blackish-brown chicks take to the water within hours of their birth and in less than two weeks are accomplished swimmers and divers. When they have had a chance to fill out a bit, their parents, sometimes accompanied by a pair of visiting loons called in from a nearby cove or lake, hold early-morning foot races across the water, evidently to train and condition the young birds for future take-offs. With wings held out and half opened, family and friends run splashing for several hundred yards, then return to the starting point, repeating the performance over and over with great zest. When the games are finally called to a halt, the racers mill around sociably in a general babble of noise in which they seem to be congratulating the young and each other as well.

In their migrations back and forth from sea to solitary lakes, loons used to present gunners with big, strong-flying and all-too-tempting targets, and although they were not generally considered as tasty as some game birds, they were heavily shot for "sport." The loon today is protected by law; the danger is no longer from guns, but from some of the very people who come to share the loon's solitude. With cottages and motorboats proliferating on more and more lakes, the remaining loons are under constant pressure to move. Occasionally you hear of someone chasing a loon around in a speedboat; I like to think that this kind of idiocy doesn't happen much. Even so, the noise of outboards on a lake can be enough to discourage nesting, and the waves they create sometimes wash the eggs out of the ground-level nests near shore. Over the long haul, two eggs a year don't leave much margin for error.

The story of the loon, with variations, is the story of many other wildlife species in New England. Much reduced in number, some are still hanging on, while others are making a surprising comeback and still others not native to the region are actually moving in. It is true that the largest predators, like the timber wolf and eastern mountain lion, have disappeared—killed off or driven north by early settlers as a threat to life and livestock. The last solid evidence of a "panther" in the region was a 182-pound specimen shot in Barnard, Vermont, in 1881. Occasional reports of sightings still persist—even as far south as Massachusetts' Quabbin Reservoir. It is possible that solitary mountain lions, as well as a few wolves, do wander down from Canada from time to time, but there are few large or remote enough tracts of unbroken wilderness to support these large, man-shy animals for very long.

The same fate nearly overtook several other species that hovered at the edge of extinction 50 to 100 years ago. By the late 19th Century so much land had been cleared for agriculture that little natural habitat remained and indiscriminate shooting, trapping and market-hunting had eliminated the caribou, and had virtually wiped out the white-tailed deer, the beaver and the moose. But as abandoned farms reverted to forest, providing cover and food, and as the dawning need for conservation resulted in stiffer laws, these animals began to multiply once more. The beaver, no longer pursued relentlessly for ladies' furs and beaver hats, is now plentiful enough to be a minor nuisance, busily building dams and flooding farmers' land. In Maine, the only New England state large and northerly enough to support a moose population of any size, the herd has increased from a remnant of 3,700 in 1951 to an estimated 15,000 today, and there is talk almost every year of having a limited open season once more.

Deer, so heavily shot for food they were almost nonexistent in places like Vermont by 1890, have flourished under hunting limits to the point that in many areas they are now tragically overcrowded in relation to the available browse of young trees and shrubs; they have become progressively smaller in size and vulnerable to massive die-offs from malnutrition and weakness during severe winters. Like the deer, grouse, woodcock and many songbirds prefer open, shrubby woods to the dense forest of the early days, and have proliferated as a result of timber-cutting and the succeeding growth of young trees and berry-bearing bushes. Even the wild turkey—the symbol of early New England Thanksgivings that Benjamin Franklin wanted to make our national bird—has been given a new lease on life: all but eliminated by the mid-1800s, it has been live-trapped from populations in New York and West Virginia and restocked to multiply into small but thriving flocks in Vermont, New Hampshire and Massachusetts. In Vermont alone, a transplanted roundup of 31 such tenacious and obviously adaptable birds has grown to a gobbling 500 or so; the toms may be seen once again—if you are lucky—strutting about and impressing the females with feather ruffs fanned out and wattles turned a furiously patriotic red, white and blue (overwhelmed females, it appears, throw themselves at their feet). No such assists from fish and game departments have been needed to encourage adaptable creatures like red foxes and raccoons, which seem to have welcomed man's intrusions, particularly his well-filled garbage cans. Going them one better are the black bears, which, in addition to ordinary household garbage, do not disdain an occasional picnicker's left-over peanut butter or tuna fish sandwich; by making do when other food is short, they have preserved their foothold in the northern woods.

These, at least, are some of the encouraging signs, though the balance is still a delicate one in the case of many species, requiring some ingenuity on the part of wildlife managers. One of their more intriguing efforts in recent years involves a creature that most New Englanders have never heard of, much less seen. This animal, called the fisher, fisher cat or black cat, is neither fisherman nor cat, but a sleek and dark-coated member of the weasel family, a big cousin of the mink and pine marten (which still exist in northern New England) and a small cousin of the wolverine (which does not). He does have the grace of a cat, but exactly why he is called a fisher remains something of a mystery: the name may have come from his supposed habit of stealing fish that early trappers used as bait, or from being confused with another weasel relative and expert fish-catcher, the otter. Most probably the name derives from his resemblance to the European polecat, a large, dark weasel also known as the fitchet or fitchew.

The fisher's special appeal, particularly to upcountry woodsmen, is that he is a bold, self-reliant loner, a far-ranging and tireless hunter of the forest. On top of that, he is extraordinarily efficient at his work: he is one of the few animals agile enough to take on the sharp-quilled porcupine, and he does so with regularity.

Because of his legendary ferocity, and the fact that he stays well hidden most of the time, the fisher is sometimes romantically pictured as a veritable giant—a black "panther" that conveniently takes the place of the vanished wolf or mountain lion in stories told around the fire. Actually, an average specimen of the fisher (*Martes pennanti*, or Pennant's marten) is little bigger than a good-sized house cat, though with his long, bushy tail a male can reach a length of three feet and a weight of 15 pounds. For sheer purposefulness and fighting skills, however, the fisher knows no match in any animal his size.

A rare daytime sight in New England is the rippling black coat of the night-hunting fisher, seen here loping off through a Vermont wood.

An eastern coyote, one of New England's newest immigrants, stands alert in the snow. Larger than his Western cousins, the New England coyote is also distinguished by probable genetic inheritances from the Ontario wolf, including a wider face, a blunter nose and a dark ruff of hair over the shoulders and down the back.

The fisher hunts mostly at night, and in winter travels alone through the deep snows of the forests of Vermont, New Hampshire and Maine. Like other predators, he is regarded by some oldtimers as a "varmint," and though he may occasionally take a tempting chicken or a winter-weakened fawn, studies show that his main diet consists of mice, squirrels, small birds, rabbits, porcupines and animal carrion. Like the weasel, skunk and other members of his family, he can emit a foul-smelling musk to discourage attackers; by mid-March, when the breeding season begins, this musk is often deposited along his trail at so-called scent posts that are thought to serve as a means of communication among mating pairs.

Once mating has taken place between fishers, an odd thing happens: after the fertilized eggs have developed into tiny embryos, they stop growing and remain dormant in the female for a full 9 or 10 months. Then, when the lengthening days of the following January and February signal the approach of spring, the embryos begin to develop until the kits are born—in late February, March and early April. Because of this strange reproductive cycle, which biologists call discontinuous development, the fisher's gestation period is 51 weeks, the longest of any North American mammal, and unusually lengthy for an animal its size. The reproductive cycle is so long, in fact, that about a week after one litter of babies is born the mother must leave them for a night or two to find a mate for next year's litter. If the male tries to follow her back to the nest in a hollow tree or rock crevice, she may chase him off savagely, apparently well aware that he has little compunction about eating his own kind.

Few humans have ever witnessed an encounter between a fisher and his pet prey, the porcupine; the combat is usually so lightning quick that those who have glimpsed it are not always sure of what they have seen. According to Dr. Malcolm Coulter of the University of Maine, a leading student of the animal, the fisher carefully avoids the porcupine's bristles, darting in to the nose and head where there are few quills. When the porky finally succumbs the fisher flips him over and opens him up like a surgeon, from head to tail. When he is finished there is only a pelt of quills, upside down in the forest, picked clean. Despite his agility, the fisher usually walks away with his share of quills sticking in him. A peppering like this would cripple any other animal, but it doesn't seem to bother the fisher. Biologists have found as many as 100 quills in a dead fisher's skin, or lodged deep in the muscles and

even the stomach. But they lie flat and harmless until they have softened and partially dissolved or have worked their way out, with no signs of inflammation or infection. It is an immunity that no one has yet satisfactorily explained.

Not only does the fisher seem to have this immunity, but his fondness for porcupine may also have been what ultimately saved him from extinction. By the early 1920s his luxurious pelt had become such a favorite for ladies' furs that a single prime skin frequently brought more than $300, the highest price of any fur-bearing animal in North America, and by the 1930s the species had all but disappeared. In the meantime porcupines in some areas had multiplied to pest proportions, gnawing their way into camps and summer homes with new-found abandon. In an attempt to bring things back into balance, New England states began to protect the fisher; Vermont actually bought 124 fishers from Maine and released them in a restocking effort to control the porcupine nuisance. The fisher is now numerous enough so that limited trapping seasons are allowed. But, as some biologists warn, the species does not reproduce rapidly and must be carefully watched. Besides his usefulness for porcupine control, they add, the "black cat" is simply an admirable animal to have around.

Many New Englanders may not have even heard of a fisher, but it has come as a shock to some of them to learn that there are coyotes in their midst. A few have doubtless muttered dire warnings about "wolves" and cast an eye at the old flintlock over the mantelpiece. More than one, imagining foul threats to livestock, has marched off into the woods with traps baited and shotgun cocked. To others, however, it has been a perfectly marvelous bit of news, the fact that the most mercilessly hounded creature in the country has been clever enough to outwit its pursuers and set up shop in Vermont, New Hampshire and Maine.

Coyotes, of course, are supposed to stay out on the Western plains, where righteous ranchers can use them for target practice. They are not so stupid. Instead they have for years been quietly doing what practically no other North American predator has been able to do in the face of advancing civilization: extending their range. In recent decades fully 19 subspecies of coyotes have been recognized by wildlife experts, spreading and adapting to all sorts of habitats from Central America to Alaska, from the forests of Ontario to the swimming-pooled canyons of the Hollywood Hills. In 1969, after much careful observation, wildlife biologists suggested that a 20th subspecies should be tentatively added to the list (with the provisional question mark cus-

A Canada jay, or whiskey jack, looks down inquisitively from a pine branch. This personable bird's instinct for collecting insects and berries—and anything he can steal from a campsite —helps him survive through long winters. The whiskey jack preserves many of his finds by balling them up in a coating of saliva, storing them in his throat pouch and then hiding them away in holes for future use.

tomary in such cases): *Canis latrans* sub(?)—the New England, or eastern, coyote.

For a long time there was great confusion in the Northeast about the unfamiliar animals that, in increasing numbers, were being shot by hunters or caught in traps. Some looked like small wolves, others like crosses between a wolf and a collie, or a basset and a coyote or something else; quite a few were what-is-its that defied any description. Many, it turned out, were wild dogs that may at one time have been household pets but had long since taken to the woods. A few were identified as "coy-dogs," crosses between some adventuresome Rover or Fido and a true coyote that had been looking for other coyotes but was willing to settle for almost any mate. Gradually, however, the pattern began to point to a distinct new variety: a more-or-less true coyote, darker-coated and heavier than his plains cousins (up to 50 or more pounds compared to 20 to 30 pounds). Over three quarters of a century, it seems, this coyote had slowly spread from Michigan and Ontario into New York and Vermont, where specimens were seen as early as the 1940s and 1950s, then into New Hampshire and finally into Maine —his larger size and color indicating that he had probably picked up some genes from the Ontario wolf.

And so the New England "wolf" had returned. With him, to the distress of many wildlife managers, came bloody cries for the traditional bounties. Patient men tried to explain that the coyote was not a "cowardly killer" but a rather admirable family man, in fact, a loyal mate and defender of his pups (who actually waves his tail slowly, much as a dog wags his, among his own kind); moreover, they said, he was a useful scavenger of the woods who cleaned up carrion before it became a health problem and helped keep rodents under control. Sure, he would take a grouse or chicken or even a weak deer on occasion; but he had no appreciable effect on the game populations so dear to the hearts —and stomachs—of hunters.

Old myths die hard. In 1973 a public hearing was held by the Maine legislature on LD 23, a bill to place a $50 bounty on the eastern coyote (of which the state had an estimated population of 500 or less). "Make no mistake about it," said one witness, who admitted he had never seen one of the creatures, "the coyote is a killer. And what if a pack of them became rabid?" "They're killing *our* deer," another hunter indignantly cried. "If you don't do something you're not going to have any rabbits, deer or moose," pleaded a third member of the audience, overlooking the fact that wolves coexisted with rabbits, deer and moose in Maine long before the white man arrived.

The wildlife professionals tried once more. "For 50 years," testified Frank Gramlich, state supervisor for the U.S. Fish and Wildlife Service, "my agency has shot, poisoned and trapped tens of thousands of coyotes. None of it has worked to limit coyote populations, except on a temporary and limited basis. The coyote has a built-in population-control sensor. When he reaches his optimum population in Maine, he will stop. Nothing we can do will change it." Another wildlife specialist pointed out that domestic dogs outnumbered coyotes 250 to 1, and ran down and killed far more deer. "People condemn the animal before they even understand him," a state wildlife biologist said.

After all the testimony had been heard, the Maine House of Representatives voted to shelve the coyote bounty bill, though by no landslide; the vote was 73 to 55.

Not all New England creatures have been under such intense pressure that laws have had to be passed, or defeated, to ensure their survival. One of the most successful species is an old and abundant north-woods resident, and although he is probably more larcenous than the coyote, he will never come to any harm. He is the Canada, or gray, jay, better known to woodsmen as the camp robber, moose bird or whiskey jack. (The latter name apparently stems from the Indian name for the bird, *wisskachon* or *wiskedjak,* though there are those who think his flight pattern is a little tipsy.) "An elegant jay," Audubon wrote of him, "joyous and lively at all times." Despite his thieving habits, the whiskey jack has long been a special friend of trappers, hunters and loggers, who still welcome his companionship, particularly in winter, when the forest can be a cold and lonely place.

If you have ever camped in northern Maine, New Hampshire or Vermont, you have probably met the whiskey jack. You don't have to call him; at the first sound of an ax or sign of campfire smoke he'll call you. He looks a little like an overgrown chickadee, soft gray with a jaunty white head, striped black on the back and masklike across the eyes. If he is especially anxious to see you, he may even perch on the bow of your canoe and watch intently until you paddle into shore for the night. More likely, though, you will first notice him as he checks out your campsite, sailing down from a nearby treetop, landing on the lower branch of another tree, hopping upward from branch to branch in an erratic spiral, then repeating this all-level reconnaissance from a closer tree. The performance may be accompanied by any number of noises —chucks, coos, trills, whistles, screams—for the Canada jay is something of a mimic and seems to enjoy showing off. If you turn your back for a minute, he will nip into your packbasket or tent, pilfering food, soap, matches, tobacco or anything else he can carry off. When you finally get supper on the fire he can become downright cheeky, watching your every mouthful, dashing in to snitch the bacon, even taking crumbs from your hand. What he cannot stuff down immediately he stashes away in the woods; if you are adept enough to follow him, you may discover several good-sized handfuls of biscuits, breadcrumbs and assorted junk stuffed into a hole in a tree. If he knows you have found his hiding place, however, he may promptly remove every bit of his treasure and cache it somewhere else.

Whiskey jacks are tough, too. Long after you and other less hardy migrants have gone south, they will remain in the north woods, building tight nests insulated with bits of hair, down and feathers against subzero temperatures and laying eggs during the snowfalls of early March. Apparently pairs mate for life, stake out a territory about a quarter of a square mile in area and seldom stray far from it all year round, defending it against gray jays but not other species. When you return next year, the jay and his family will be there to greet you and make genial pests of themselves again. Old loggers and woodsmen may tell you, half seriously, that these birds are the reincarnated souls of lumberjacks, and that great evil will befall you if you ever do a whiskey jack harm. But, really now, who could hurt an old friend?

A Fledgling Generation

The return of warm weather to New England is a time of renewal for all things, including the region's teeming population of birds. From farms and pastures to the wilder mountain forests, an estimated 100 million birds fly about establishing territories, seeking mates, building new nests or repairing old ones, all bent on the serious business of perpetuating their kind.

Evolution has equipped different species with different ways of maintaining their numbers in the face of predators, bad weather and other factors threatening the lives of their young. Even the best-built nest can collapse in a hailstorm, its eggs smashed to the ground, but if this happens or if the eggs are stolen from the nest, the females of most species will usually attempt a second nesting, though often producing a smaller second clutch. Protective coloration also helps some brooding birds such as the ruffed grouse at right, whose mottled brown plumage blends well with the leaf-covered forest floor; many smaller birds, like the yellow warblers seen overleaf, seek nesting cover in dense vegetation that discourages intruders.

Predators are frequently held at bay more by ruse than by strength. The mother grouse is famous for her trick of feigning a broken wing and flapping around, apparently helpless, to divert attention from her chicks. Young ospreys, responding to a voice signal from a parent, disappear below the rim of the nest and play dead. If it comes to a fight, however, birds like goshawks fiercely drive off crows or other hawks, and even flightless young owls defend themselves with beak and claw.

Nestlings of all species are constantly protected by their parents. Warbler chicks are fed frequently and increase their weight several times over within their first week. Female great horned owls, which nest in New England as early as February and March, continue brooding their young even when lightly covered with late-winter snow.

When the time comes for the offspring to try their wings, the adults have various ways of teaching them: ospreys, for example, may tempt their young by holding fish at a distance to urge them to fly a short way from the nest. Finally, when the fledglings have gone off on their own, the ultimate goal of the breeding season has been achieved, and the survival of the species has been assured for another year.

Less than a day old, a ruffed grouse nestling snuggles under the wing of its mother. Hatched from a clutch of 8 to 14 buff-colored eggs, the young walk minutes after they are born and are able to fly in about 10 days.

A yellow warbler chick nested in a flowering wild rosebush presents the red target of its gullet to parental feeding. Both adults bring food to the nest every few minutes to satisfy their offsprings' lusty appetites.

Firmly holding the body of a chipmunk, a goshawk feeds bits of meat to two snowy nestlings. This powerful and handsome bird preys mainly on grouse, rabbits and other small mammals.

Young ospreys practice their flight exercises at the edge of the nest. Ospreys' nests, sturdily built of sticks and other plant debris, are strong enough to support a man's weight.

A great horned owlet is startled after a fall from its nest. Its parents will continue to feed the owlet on the ground and in the low branches where it may climb for greater safety.

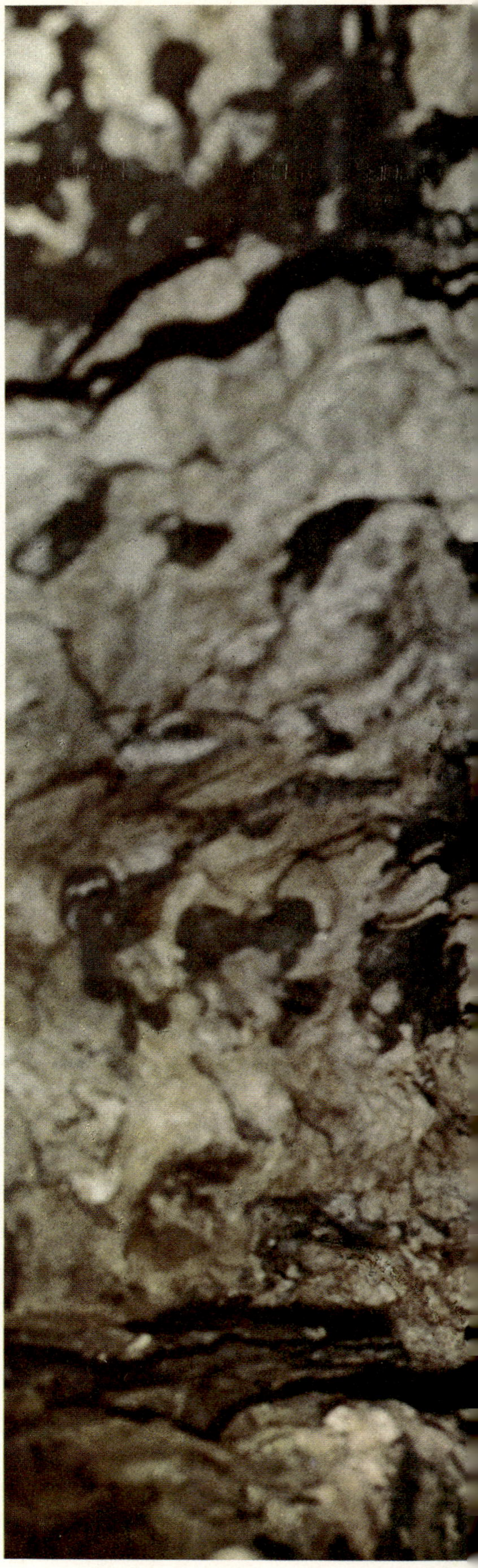

Three young barred owls peer at their surroundings from the shelter of an old maple tree. Like their parents, they have the extraordinary ability to swivel their heads to almost any angle.

5/ A Private Wilderness

A lake is the landscape's most beautiful and expressive feature. It is earth's eye, looking into which the beholder measures the depth of his own nature. HENRY DAVID THOREAU/ *WALDEN*

Behind the granite curtain of the White Mountains, beyond the public playgrounds that magnetize visitors from New Jersey and New York, lies a swatch of territory that some New Englanders would just as soon keep for themselves. Loosely referred to as the "north country," it is not so much a geographic region as it is a frame of mind, a last frontier against development from the south. It has no precise boundaries, but meanders across northern Vermont, northern New Hampshire and on up into northwestern Maine, broken only by a few principal roads. Along the way it embraces the headwaters of the Connecticut, the longest river wholly within New England, which rises in a lonely chain of lakes near the Canadian border and winds through picture-book farms and forests before turning south to Long Island Sound. The north country is also the birthplace of other rivers like the Androscoggin, which flows from Lake Umbagog, taking in the ambling Magalloway River, then loops through the belching paper-mill towns of Berlin, New Hampshire, and Rumford, Maine, before it joins the Kennebec at the sea.

The north country is logging country, as it has been for two centuries or more, blanketed in maple, beech and birch, spruce and fir. It is wild, wet country, dotted by swamplands, ponds and lakes, its tangle of undulating hills rarely rising over 3,000 feet but occasionally producing some spectacular notches and ravines. Except for a few patches here and there, it is all privately owned, mostly by big companies like St.

Regis, Scott, Brown, Great Northern, and International Paper, who manage an intricate mosaic of holdings for the production of pulpwood and lumber, and who keep their lands open to outdoorsmen, except where the woods are being actively logged. It is country that still boasts some of the best hunting, fishing and wilderness hiking in the Northeast, for those who know where to go—and how to get back out.

One of the guardians of the north country is a cheerful man called Smitty—his full name is Robert Smith—a former appliance and soft-drink salesman who decided to go back to the woods, and became a fish and game warden 12 years ago. Smitty's territory, which happens to be in the very northeastern corner of Vermont, takes in 237,000 acres and is among the larger such districts in New England. He covers his beat pretty much by himself, driving around backcountry dirt and gravel roads in a green International Harvester pickup truck equipped with a two-way radio, a .30-.30 rifle on the dashboard and a square-ended canoe lashed on top. He spends a good deal of his time tracking down illegal deer kills, pulling capsized motorboaters out of lakes, and finding hunters and fishermen who have become disoriented in the woods. He knows roughly where he can find a dozen moose and the dens of five coyote families, and he has plenty of reports of bobcats, black bear and even marten and Canada lynx. Deep in the woods he has seen canine tracks four inches across, as big as those of his 130-pound bloodhound, though like many north-country residents he hesitates to say flat out that it's a timber wolf. (A few years back he got a report of a walrus in the Connecticut River, and passed it on to headquarters, much to his regret; the dispatcher still asks him from time to time, "Say, Smitty, how're the walrus running up there this spring?") He is, altogether, quite proud of his kingdom, as he calls it, and he likes to show visitors around. He is proud of his uniform, too, although he often finds it just as comfortable to show up, as he did not long ago, in a baseball cap and an old blue and white team jersey.

The first corner of his kingdom Smitty showed me was one that has achieved some fame in recent years. Like so many now-deserted communities in New England, the town of Victory was once a thriving settlement; in its heyday more than 500 people lived in the area, "letting daylight into the swamp," as the old logging expression goes, as well as farming the cleared-off, shallow soils. The logs were brought out on a little branch railroad built for the purpose, cut up in a dozen sawmills locally or floated on down the Moose River and the Passumpsic to the Connecticut for the big spring drives *(picture essay following)*. By the early 1900s, the virgin growth was almost gone and the soil was giving out; though some cutting still went on, the railroad branch had pretty much had it and the last rails were pulled up for scrap in 1917.

Today the population of Victory is down to fewer than 50 people, scattered along one or two dirt roads, but the name has taken on something of a new ring. As we slogged in that morning from Damon's Crossing, I found it difficult to believe that nature could have taken over so completely again. The old rail bed, now used as a trail, was barely discernible, rank with head-high alder bushes and flooded here and there by the outwash from beaver ponds. On higher ground the forest had closed in thickly, spruces and firs growing in impenetrable ranks hardly more than a foot apart. At the base of Bog Pond, one terminus of the old railroad spur, a broken dam and a couple of overgrown cellar holes were the only visible evidence of a hamlet that had once prided itself on 20 houses, a store, a school, and a lumber mill that consumed five million board feet of spruce logs a year.

The serene, shallow waters of Lake Umbagog, on the Maine-New Hampshire border, adjoin marshes where the Magalloway River ends and the Androscoggin begins. Umbagog's numerous coves and inlets, and the wetlands that surround them, comprise one of the finest wildlife areas in the northern part of New England.

What makes Victory particularly sweet to Smitty, however, and to many other inhabitants of the region, is that they were able to snatch it from the jaws of developers—in this case the Army Corps of Engineers. The corps wanted to put a five-million-dollar flood-control dam across the Moose and turn the whole rich complex of forest and wetlands into one great big "recreational" lake, full of warm-water fish and speedboats, with a thousand campsites around the shore. It took some standing up to the federals, as well as to a senator and a governor, plus the campaigning of a local museum director named Fred Mold, and an Audubon Society report in which a handful of professors and interested amateurs showed that the dam would be not only an ecological disaster but dam foolishness for flood control as well; the spongy wetlands act quite nicely as their own regulator of flow. Victory, at least 5,000 acres of it, is now owned outright by Smitty's fish and game department as a wildlife management preserve; timber-cutting is controlled to provide maximum cover and food for deer, grouse and other species; hiking and canoeing are encouraged but overnight camping and the use of any motorized vehicles, including snowmobiles and trail bikes, are banned. You can still go to Victory and hunt deer or catch brook trout, or you can just look at the showcase of life: quaking bogs where many varieties of northern orchids grow; open meadows bright with blue flag iris; small dry knolls covered with partridgeberry, dogberry and pink lady's-slipper; and animals ranging from muskrats and

otters to ospreys, marsh hawks and rough-legged hawks wheeling overhead in the sky.

That day Smitty also took me to another corner of his kingdom, a sort of north-country Walden still unspoiled (the original Walden, in Concord, Massachusetts, is now a tourist mecca and state park). I had spotted it on a topographic map, looking scarcely larger than a raindrop splashed down in a puddle of brown contour lines at 1,667 feet. In shape it was almost a perfect ellipse, a tiny blue eye in the background of forest green. Its name was Unknown Pond.

Perhaps I should say at the outset that there are several Unknown Ponds in the north country—not to mention an Enchanted Pond and an Ugh Lake. None are particularly easy to get to; in fact, I might have trouble finding this one again myself. In any case, as I discovered when I got there, it is someone else's Walden, its single campsite leased from a paper company, and the owner would probably like to keep it that way.

To get to the pond, I drove behind Smitty's pickup 20 miles north on a highway, onto a side road that followed a pretty stream, then up a dirt path so full of potholes and washouts that I expected several times to hear an axle snap. After six or eight miles of jouncing and weaving, Smitty finally pulled over to the side of the road. "Trail's right in there," he said cheerfully, pointing to a grim, unbroken wall of woods. At this moment a strange thing happened: a nice-looking man and woman with packbaskets strapped to their backs emerged from the trees. Stranger still, after chatting with us for a while, they told me where I could find the key to their camp. "You might like to get in away from the black flies," the man said with a smile. "Besides, it looks like rain tonight. There's split wood on the porch, and a bag of birch bark to start it inside. Just lock up when you leave." A little taken aback by this turn of events, I thanked them and waved goodbye.

The trail, when Smitty found it for me, led up through deep, wet, winding woods for a couple of miles, then abruptly debouched at the pond. Sloshing through an alder thicket, we climbed to a small, neatly built log cabin set back among the trees, and sat awhile talking and looking at the view. Then Smitty, having delivered me safely, told me to enjoy myself and headed back down the trail.

That evening, after I found that the fish weren't biting, I was happy just to settle with my back against a porch post and watch dusk come over the pond. The oval of water was as perfect as it had been on the map, less than half a mile long and a few hundred yards wide, about the right size for one man's lake. Cupped in a hollow of forested hills, it mirrored a dark, delicate eyelash fringe of spruces and firs on the opposite shore. A thousand feet above rose an irregular mountain, a lower outlying peak jutting up near the pond and dipping back to a higher, more distant one. A low batting of clouds hung on the summits, trailing wisps of fog down their sides. I could barely make out a tiny object on the topmost ridgeline: an old radar station I had heard about, built years ago in this remote country as part of an early-warning system, a military ghost long since stripped of its expensive hardware and abandoned to the porcupines.

Down in the thicket that marked the inlet of the pond, a pair of red-winged blackbirds, their gold-fringed epaulets flashing, chucked and twitted busily, sometimes flying up to a nearby branch to inspect me, then dashing away again to chase each other through the brush. A flurry of little swallows pinwheeled above the water, fluttering one minute like a handful of confetti, dipping and soaring the next like paper airplanes. As the light faded, the mutterings of frogs built to a chorus of a thousand singing voices, pierced now and then by the liquid soprano of a thrush or the drum of a woodpecker carpentering a tree. By 8:30 the palette of greens on the hills had merged to a hazy purple; the mountain was a gray cardboard cutout in the sky. An evening wind rose in the firs and spruces and their images blurred as a brush of evening rain stippled the surface of the pond. At 9:30 the light was almost gone. I was thinking about Smitty's coyote families in the hills when a weird, chilling cry came from the mountainside: four measured *hoo*-ing calls, the last two linked in a throaty, trailing-off sound. It might have been a coyote. Then again, it might have been an owl. With the rain beginning to pummel down, I was thankful to retreat to the cabin, and unrolled my sleeping bag as the fire crackled itself to sleep in the stove.

Morning at Unknown Pond brought no great burst of bird song, no brilliant sunrise, just more of that abundant New England commodity, rain. Packing up reluctantly, I set out to explore the forest and return by another route the owners had described.

The woods of the north country, while you're in them, are a fine place to get to know a variety of trees. Besides the Christmas-tree profiles of red, black and white spruce and balsam fir, there were the flat, outsweeping sprays of hemlocks and the slender wands of larches by the swampy outlet of the pond. The eastern larch, often called the tamarack or hackmatack in these parts, is the most distinctive of the region's cone-bearing trees: along with the bald cypress of Southern swamps it is the only "evergreen" to lose all its needles each fall; all winter its tall, straight trunk and prickly-stubbed bare branches make it look like a forest-fire victim, until its bright yellow flowers and its whirls of feathery, light-green needles return in spring. On drier ground were sugar maples, surely the best loved of New England's trees —graceful in their upward-reaching symmetry, masterful in their geometric detail of five-pointed leaves and winged seeds, sweet of sap, a glorious red and orange in autumn, their wood among the most prized for flooring and furniture as well as for fires. Here and there were other members of the northern hardwood forest: red maples, whose explosion of pink-red flowers, opening before the leaves, are a trademark of early spring; handsome American beeches with their tight, smooth trunks of aluminum gray; big yellow birches, their bark gently frizzled with golden-silver curls. Below these were the smaller trees of the forest understory, notably the delicate, white-flowering shadbush, or serviceberry, and the striped maple, or moosewood, with its large three-pointed leaves and satiny green, white-streaked bark—a favorite food, along with the tender buds, of deer and moose. And finally, in a clearing, a band of young white birches, the true ballerinas of the forest, arching and shimmering against the dark backdrop of the woods.

At this point, in the company of trees, I experienced a new and not altogether welcome sensation: I was not sure where I was. What had seemed a trail was no longer to be seen. I struck out in one direction, hoping to intercept it again. No luck. I returned to the brook, thinking to follow the water downhill and out. The brook, perversely, became two or three brooks, braiding out through woods and marshy depressions, appearing and disappearing as if to tempt me on. I cut back toward higher ground once more, stumbling through bushes and over rocks; a root hidden under a mushy mat of leaves sent me suddenly flying into a water-filled hole, where I landed on my backpack like a jackknifed trailer truck. Struggling up, I cut to the right again, found what looked like an old logging road and followed it downhill for a while, only to see it peter out in a cut-over clearing. Here the logging swaths, growing up to a pale green jungle of vines and shrubs, led out in all directions toward darker jungles beyond.

It hadn't occurred to me that people actually got *lost* in New England. But as I later found out, they do, and with surprising regularity. City dwellers who wouldn't dream of exploring Alaska or the Rockies without a guide and a truckload of the latest survival gear take off into

An American toad peers impassively through

a curling blade of grass. Much in evidence in New England in summer, this toad spends the winter dormant in burrows or under rocks.

the backcountry of Vermont, New Hampshire and Maine with their street shoes, a chocolate bar, and little else. Contributing to their euphoria, no doubt, is the sight of open farmland, roads and houses, often right up to the point where they decide to leave their cars and set out on foot. Once in the woods, however, the bucolic scene can change with unexpected swiftness, particularly if the visitors leave established trails and start off cross-country to admire the flowers or track a deer. Suddenly one tree can look very much like the next, and what landmarks they might be able to see and recognize through the forest—a mountaintop, a pond, even the sun—can disappear in fog, rain or snow. Notions dimly remembered from Boy Scout days, like following a stream, can get them into even more remote areas and worse trouble than they were in before. Many stubbornly keep on wandering, instead of giving up, staying put and waiting for a search. When the truth finally sinks in, human reactions can complicate matters still further: some people become actively nauseated, while others start imagining things. As Smitty told me that morning, he once rescued a man who was a few hundred feet from a long-deserted logging road and heading resolutely toward it, convinced he could see the lights of cars and trucks. Like other poor souls rescuers have found in the woods, he couldn't quite bring himself to admit being lost, just "turned around a bit."

Well, I was turned around a bit, all right. I remembered a pamphlet I had stuffed into my pack, one of these cautionary tracts prepared by a paper company for people who use the woods. Fishing it out, I peered at the lettering through the rain. *Notify a game warden of your planned destination. He is your friend and a professional woodsman, trained to protect and assist you.* Check. I have a friend at Fish and Game, I think. I said if I hadn't called him by 4:30 this afternoon would he please, as a favor to my wife and children, send out the National Guard. *We suggest you take along a topographic map and a good compass.* Check. On my map I should be somewhere north of the pond, which means if I head east I should eventually hit the road, somewhere. In this particular area the map appears liberally sprinkled with blue, tuftlike symbols indicating bogs or swamps. Apparently heavy going, if not outright quicksand. But is a 98-cent compass a *good* compass? A little late to be considering that. Just put it on the ground, away from pocketknives, dental fillings and other metal objects, and see if the needle points somewhere. Map indicates I should add (subtract?) 17° for 1956 magnetic declination from true north at center of sheet. I am not at the center of the sheet, and this is not 1956, but I suppose that can't be helped. Raining buckets now. *If you become lost, DO NOT PANIC.* Who, me? Smitty told another little story yesterday about two girls who got lost only a quarter mile from a road; they could hear their rescuers looking for them most of the night, but they were throwing up so much out of sheer fright they couldn't utter a sound. *Tramp out signals in the snow: I for I require doctor, X for unable to proceed, F for need food and water.* There is no snow. Why should there be? It is summer. Not much food either, come to think of it, for spending a night in the swamp. *Carry a small mirror or other shiny metal object to signal search aircraft.* Who in his right mind would be flying on a day like this?

When the compass needle finally stopped shaking I made a crude estimate of east and pushed on through the brush. Now I know why they call it hobblebush: let a little daylight into this swamp and it sprouts barbed wire. Check east again: it seems to have gotten lost. Up over that rise? A problem. Either I have arrived at the blue-tuft symbols, or the beaver have been at work: a veritable slough of despond glitters evilly ahead, left and right as far as the eye can see. Only one thing to do, cross it, up to your wallet in water. No quicksand, luckily, but this water must be about 30°. Now get over that maze of fallen trunks and up to the top of the hill. What is that thin white line glinting through the trees, another swamp? No, DO NOT PANIC—IT IS A ROAD!

On the road I headed south, and after a mile or so, oddly enough, I came to my Volkswagen, just where I had parked it what seemed a few months ago. Driving out on Axlebreaker Boulevard, I found a phone on the highway and called Smitty at 4:25 through the State Police; while I waited, a trooper on the other end politely relayed the message to the radio in his truck. All I could hear the warden say over the static was some matter-of-fact code acknowledgment like "Ten-Four." Just another routine day in the woods, I guess. Except for a cheap compass, Victory and a pond I know.

The Bangor Tigers

The woods of northern Maine, New Hampshire and Vermont still echo with the legends of the old-time loggers, a tough, colorful and virtually bygone breed of men. During the last century and into the early 1900s they made New England the nation's premier lumber-producing region; for a time the world's leading timber port was the city of Bangor, near the mouth of Maine's Penobscot River. Of all the men who worked in the forest—a polyglot mix of Yankee farm boys, French Canadians and Indians, with a scattering of Irishmen, Swedes, Poles and Finns—the elite of logging were the Penobscot men, or Bangor Tigers, as they called themselves—a name soon adopted by other loggers around Maine. Skilled woodsmen, hard to beat at dam-building, boat-handling and saloon fighting, they were amazingly nimble—"catty" was their word for it—riding a bucking, rolling log down a spring-flooded river in their heavy spiked boots.

The north-woods wilderness was home for these men and they sought it out, learning its ways, moving ever deeper into the forest. To stock-pile the long logs for the spring drives, they chopped away through the winter; they sledded the cut timber to the riverbanks on well-iced roads, using horse teams to pull the heavy loads. They worked in temperatures of 30 below in January and February, and stood in icy water up to their waists in March and April, relying on keen reflexes to escape falling trees and branches, runaway log loads and white-water jams. Far from doctors, they treated their minor wounds with salt pork or tobacco juice and sewed up frost-cracked hands and feet with needle and thread; if a man felt sluggish, he downed a few spoonfuls of kerosene to "tone up the system."

Yet, for all its hardships, low pay, long hours and miserable living conditions, logging was an exhilarating life, one that made ordinary jobs drab by comparison. Every summer as the men ended the drive and headed for Bangor, many a logger swore he would never go back again. But after he had drunk up his pay and seen the sights, more likely than not he would look over the recruiting lists, pick a camp with the best boss and cook he could find, and sign up for another season in the woods.

The equipment of three choppers photographed in 1909 illustrate changes overtaking their craft. The single-bladed poleax, held by the man at center, was yielding to the double-bitted ax, at right, which could be sharpened on one side with a thin, keen blade for hewing tree trunks and on the other with a thicker blade for lopping off branches. Some proud choppers scorned the crosscut saw, on the left; though two men could cut twice as much wood with a saw, felling trees with an ax required real skill.

Inside a typical logging shanty in backwoods Maine, the camp boss, eating an apple, plays checkers while some of his crew look on.

Long Winter Months in a Wilderness Camp

From October to March, while the loggers worked in the forest, they lived in crude shanties constructed of spruce logs, floored by earth and heated by a fire pit under a hole in the roof. The roof itself was built of shingles held down loosely by poles so the men could push it up and escape in case of fire. To economize on space and warmth, the loggers slept together on a bed of boughs covered by blankets sewn together into a huge communal quilt. Later camps *(left)* had the luxuries of double-decker bunks and cast-iron stoves.

Every morning at about 4 o'clock the men were wakened by the cook banging on an iron pot. Breakfast, the first of four meals, was usually salt pork and baked beans, biscuits, molasses and scalding black tea. Twelve or 14 hours later the men returned from the forest for the day's last meal—much the same as the first. After an evening pipe and a game of cards, wet socks and boots were hung by the fire; before 9 o'clock all had crawled into bed.

Sundays the men got up later, wrote letters, greased boots, sewed up clothes or looked at old *Police Gazettes*. Some went out to bring in fresh deer or moose meat, which they called Canada beef. Others shaved or boiled their clothes to kill the lice; almost none bathed, except for a few Finns who sweated awhile in crude saunas they had built.

Carrying yoked buckets of baked beans and tin teapots, dippers and cans of molasses, a trio of young cook's helpers, or "cookees," heads out with one of the two lunches taken at 10 a.m. and 2 p.m. daily to the loggers busy in the woods. The tea, which the men claimed was "strong enough to float an ax," usually arrived hot and the cookees would build a little fire to warm up the other provisions. Then they would run back to camp to do the dishes, stoke up the stove and help the cook prepare the next meal.

Alternating strokes, axmen cut a big birch near Mount Katahdin around 1900.

Hard Work in a Frozen Forest

At 5 a.m., or as soon as enough light had penetrated the frosted winter woods, the cutting started. Each chopper notched a tree deftly, then was joined by a second chopper who whaled away on the opposite side. Sometimes the felling was done so meticulously that the tree would squarely hit a stake set in the ground as a target—and as a way to show off. But the tree might also split or topple in the wrong direction, or a dead branch—called a widow-maker—could slip lethally from on high. That night in the shanty each man would put a week's pay into a battered hat for the widow; there were no other death benefits.

As the thermometer dipped near zero, the men started hauling the long logs on big sleds to riverbank landings to await the spring drive. The sleds traveled over roads carefully built the summer before. On the coldest nights, while others were asleep, two men took out a sled with a 500-gallon tank that sprinkled water on the roadway, turning it into a sheet of ice.

Transporting the timber was perilous work. During the loading, logs could easily roll off the sleds and crush or maim a man. Moving downhill on the iced roads, the heavy loads often got out of control, sweeping the horses off their feet and piling logs, animals, sleds and teamsters into a tangled, broken heap.

In a photograph taken near the Mattawamkeag River, a pair of loaded sleds prepares to move out on the icy haul to the riverbank.

White water diverted from Shin Brook boils down a sluice built to carry logs around a 40-foot-high falls, preventing the timber from piling up at the bottom. The sluice—built in two sections with a bend in the middle where a man was stationed to "tend out," or prod, balky timbers—brought the logs down to water fast enough to keep them moving downstream.

Dangerous Days on the River

The instant the ice broke up in the streams, in late March or early April, the men swarmed to the landings and began rolling the logs into the waters that would carry them to the mills and ports. It was a race against time: the rivers, fed by melting snow, had to move the logs before the water dropped. To shorten the odds, devices such as sluices *(left)* and "squirt dams" were built. The dams held back water at times of peak flow; later, the loggers would open the dams from time to time to give the dwindling streams a squirt.

It was the skill of the men that won this annual race. Wading into the icy water, they would stay there, on and off, for two or three months, prodding logs loose while their feet swelled and cracked. There were few Sundays off, no dry clothes, only labor from dawn to dark.

The river drive was the logger's supreme test. The heavy, slippery logs often jammed up behind rocks, against each other, against the shore, creaking and groaning in monstrous tangles of wood. The crew's best men were those assigned to clear the jams, working from boats with poles, leaping catfooted in spiked boots from log to log. At the very moment of success they were threatened by the unleashed timber; many a pair of boots hung on a riverbank tree signaled the death of a Bangor Tiger overtaken by the logs.

Clearing a landing, men roll·logs into Wassataquoik Stream using peaveys, spiked poles with hinged hooks that were invented in Maine.

A double-ended bateau, responsive in rough waters, waits while nine men try to pick apart a jam on the fast-flowing Wassataquoik. The instant the jam "hauled," or broke, the logs would shoot out erratically, and the men's lives depended on getting away fast. It was a dangerous maneuver and the boatmen had to be virtuosos; overturned bateaux killed more men on the drive than any other single cause.

6/ The Maine Woods

The scenery... about the head waters is grand and picturesque. Its numerous water-falls, its swelling... mountains, and the vast groves of towering pines make it altogether one of the wildest and most romantic portions of country. JOHN S. SPRINGER/ *FOREST LIFE AND FOREST TREES*, 1851

Soon after it crosses New Hampshire's eastern border, the Appalachian Trail dips sharply from Goose Eye and Fulling Mill mountains, and plunges into the dank embrace of Mahoosuc Notch. It is a wild, weirdly beautiful place, to some veteran hikers the ruggedest and remotest mile on the trail. Here the valley walls, studded with spruces clinging to the rock, are so sheer and close-set that the sun reaches in only briefly at midday; a backpacker, often enshrouded in swirling mist, climbs over, around and under giant boulders, slipping and scrambling on wet, moss-covered rocks, passing caves where ice still glistens in midsummer. For the comparatively few people who have traveled the notch—and more than one has been carried out with a painfully twisted ankle or knee—it is a rough and fitting introduction to what Thoreau called Maine's "damp and intricate" wilds.

Maine's wilderness, in fact, was sufficiently damp, intricate and rugged that the trail's early planners almost gave up the notion of penetrating it and were about to make Mount Washington the northern terminus. But a study group finally pieced together a route, and the last two-mile link was cut in 1937 by a CCC crew on the north slope of Mount Spaulding, south of Sugarloaf. If you do enter Maine on the trail, it is from these summits—Old Speck, Elephant, Saddleback, Spaulding, Sugarloaf, Bigelow—that you become most dramatically aware of the scale and character of the state. By New England standards, at least, Maine is big, nearly as big as its five sister states combined—its northernmost and wildest county, Aroostook, is larger than Connecticut and Rhode Island put together. The state is heavily wooded, the most heavily, in fact, of any in the union—close to 90 per cent forest. And it has abundant water: its 2,500 lakes and 5,000 streams make up almost a tenth of its 33,000-square-mile area and stand as the freshest and largest remnants of the glacial blankets that once scoured and reshaped the Northeast. From the summits, the panorama of these lakes unfolds: Aziscohos, the Richardsons, Mooselookmeguntic, Rangeley, Flagstaff and, to the northeast, Moosehead, largest of them all.

The view from the bare, symmetrical cone of Sugarloaf is spectacular enough. Even better is the outlook from the Bigelow Range just to the north: from the firetower on Avery Peak you can see lakes, mountains and rivers clear into Canada. If you have a mind to, you can also trace from here a good section of one of the most celebrated wilderness journeys in American history: Benedict Arnold's daring, disastrous march north to attack Quebec in the fall of 1775. From the Kennebec River, visible to the east, Arnold's army of more than a thousand men lugged their leaking bateaux and supplies, crossing the three Carry Ponds near the present route of the Appalachian Trail. They struggled up the Dead River Valley, now dammed and inundated to form the 26 miles of Flagstaff Lake directly below you, then up the river's north branch into Canada, far off to the left. It was a tragic masterpiece of bad timing, bad maps and bad luck. After two months of dragging their boats through icy rapids and tramping through woods and swamps, only 600 or so ragged, exhausted soldiers made the St. Lawrence; and when they finally attacked the city on the night of December 31, most of them were captured or killed.

From Bigelow, the trail descends and weaves around the Carry Ponds, crosses the Kennebec, then breaches the Barren-Chairback Range to enter the lake and river country of central Maine. One of the most fascinating side trips in this region is a day's hike up a blue-blazed trail into Gulf Hagas, a remote and wild place where the West Branch of the Pleasant River cuts its way through a slate canyon, dropping 500 feet in less than three miles in a series of spectacular cascades and falls.

While the Appalachian Trail is still a good 40 or more wet, winding miles from its end, you begin to become aware of its terminus: the great, brooding presence of Mount Katahdin, actually a whole range of peaks just north of the center of Maine that dominates the lake country in all directions. Katahdin, the guide books tell you, is the first piece of United States real estate touched each morning by the rising spring sun. At 5,267 feet, its summit is not very high above sea level, but the way the mountain rises abruptly from the forest in splendid, distant isolation, once seen, is something not likely to be forgotten.

To the Abnaki Indians who traveled the woods and waters beneath it, Katahdin—which means, roughly, greatest mountain—conjured up potent visions. One of them was Pamola, the "storm bird" of Indian mythology that had the wings and claws of an eagle, the arms and torso of a man, and the head and antlers of a moose; when aroused, this fearsome triple threat unleashed forked lightning, snowstorms and violent winds to prevent men from climbing his sacred peak. Indian braves foolhardy enough to venture above the trees were struck with delirium, or woke up at the foot of the mountain, or vanished completely into the air. The Pamola vision probably suited a lot of Indians, who were too busy making a living in the woods even to think about mountain climbing, particularly in bad weather.

If the Indians considered mountaineering profitless, they took great pride in their canoeing, and when the white men came to explore and exploit their wilderness, they were hired on as rivermen and guides. By the late 19th Century the West Branch of the Penobscot, at the foot of Katahdin, had become the scene of Maine's biggest and most colorful logging drives—a veritable river of wood that every spring ran all the way to Bangor, where lumber ships destined for ports around the world were lined up so thick, the boast went, you could cross the harbor on their decks. Boss of the West Branch drive was John Ross, whose Bangor Tigers looked upon him as the power that commanded wind, water and logs. One of the Tigers' favorite stories was about the day that a famous Indian riverman, Big Sebattis Mitchell, became the first, and the last, man ever to run Nesowadnehunk Falls.

To get up and down the river, to haul supplies and tend log jams, the loggers used big two-ended bateaux much like the ones Arnold had taken on his journey a hundred years earlier. Carrying the heavy boats around unrunnable sections of the river was sweaty, backbreaking work. It seemed especially so one warm May day in 1870, when two crews of river drivers had hauled their craft around the falls and were waiting for the third. "Big Sebat," a member of the Penobscot tribe and a bear of a man at 260 pounds, sat for a while in the stern of the third boat, his paddle across his knees, talking with his bowman. Apparently he had had enough portages for the day.

Suddenly one of the white loggers lounging in the shade beneath the falls glanced up and gasped. The last bateau seemed to pause for a second, framed on the lip of the upper falls. Then it shot clear and, flying out, landed in the boiling rapids 8 or 10 feet below. Both Indians in the boat leaped to their feet, whipping their paddles from side to side to avoid the litter of rocks and logs. At length they fetched their half-swamped craft ashore in a cove around a bend, rested for a minute, then bailed her out—all but a couple of pailfuls, for appearance' sake. When their incredulous companions hurried up, there stood Sebat and his bowman, leaning on their paddles like a pair of smiling bronze statues. It was more than the pride of the other rivermen could stand. Without a word, they walked back to their own boats, lifted them onto their raw shoulders and started back up the portage trail. They too ran "Sowdyhunk" falls—mostly upside down. All but one man were rescued; the boats were a total loss. When Big Sebat used to tell the story there was one part he couldn't get over: John Ross didn't seem to care so much that the stunt had drowned a man—there were plenty of loggers looking for jobs—but he gave Sebat hell about the smashed boats.

From just below Nesowadnehunk Falls, the Appalachian Trail begins its ascent to the base of Katahdin, winding up Nesowadnehunk Stream, where the logs once rolled and banged over foaming cascades called Little Niagara and Windy Pitch. Here the path crosses an old logging tote road, now the main road into the area from the south, and enters Katahdin Stream campground, the last stopping place before the climb to the top.

The final leg of the trail is only a little over five miles long, but it rises a stiff 4,000 feet, and you have to push to get over the boulders on Hunt Spur. ("A vast aggregation of loose rocks," observed Thoreau, "as if some time it had rained rocks, and they lay as they fell on the mountain sides, nowhere fairly at rest.") About a mile from the top you reach the high, barren plateau called the Tableland, then press on through blowing clouds to the summit. There is a sign there, pointing back 2,000 miles to Georgia, and a bronze registration cylinder where you can leave your name. There is also a view second to none in the Northeast, of the forest and a hundred mirrored lakes. Immediately to the east is the Knife Edge, a razor-like, mile-long ridge that forms the rim of the Great Basin, a giant amphitheater cut by glacial ice. It is no place for acrophobes; in spots it seems every bit as narrow as its name suggests, and when the fog is swirling you may have to crawl it on your hands and knees. Off to the west is the windswept saddle of the Tableland. Directly below, another trail dives down over rubbled slopes to Chimney Pond, a clear, frigid glacial tarn where hardier mountaineers camp for winter as well as summer assaults.

Spreading out beyond these landmarks is Katahdin's larger setting, the 200,000 acres of Baxter State Park. If there is a swatch of purest gold in the patchwork fabric of New England's wilds, this is it. Baxter's solitude and scenery have captivated almost everyone who has been there, from Thoreau and other early travelers to Justice William O. Douglas, who has been drawn back again and again by the "haunting melody" of the place and who has found it, as I do, at its greatest glory in fall. It was also beloved by the late Percival Proctor Baxter, and the result of his love affair is a story unique in American conservation.

As early as 1860 a few lonely voices were suggesting that the Katahdin area be rescued from the loggers and land speculators; in 1916 a Maine Representative tried unsuccessfully to get Congress to make it a national park. By then Percy Baxter, a lawyer and Maine state legislator and the son of a wealthy Portland businessman, had become entranced by the beauty of the region. His early bills to save Katahdin as a state park were twice turned aside. In 1920 he tried again, this time appealing to local pride by asking that it be made Centennial Park in celebration of a century of statehood. No takers. A general apathy, coupled with the opposition of land and timber interests, again proved too great, although Baxter was able to get the legislature to create a game preserve of 90,000 acres on and around the mountain. He continued to pursue his goal through two terms as governor, speaking up and down the state to anyone who would listen. "I was attacked as a dreamer and branded as a socialist," he recalled.

Baxter could be as stubborn as any Yankee, but in Katahdin he ran up against a granite wall. So he did the only thing a good American capitalist could do: he bought the property. By 1930 he had quietly persuaded the Great Northern Paper Company to sell him 6,000 acres, including most of the main mountain, and he deeded this parcel over to the state; the legislature accepted, and was even gracious enough to change the name of the highest summit from Monument to Baxter Peak. In 1939 he bought and turned over another 12,000 acres, and in 1940 and 1941 he gave 16,000 more acres, carefully wrapping up each gift with legal ribbons that left no doubt as to his intent: to keep the land forever wild. Well before his death in 1969, the scope of Baxter's life

Sunlight glows on Billings Falls at the head

of the three-mile gorge of Gulf Hagas in central Maine. The golden reflections are limonite, a product of iron oxidizing in the rocks.

work, as he referred to it, had become clear: he had singlehandedly purchased, protected and even partially endowed with operating funds no less than 201,018 acres of wilderness as a public preserve. At one point there was pressure to turn Katahdin over to the federal government as a national park. Baxter, fearing the crowds and commerce that might follow, vehemently refused; who needed Washington anyway? "Katahdin in its grandeur," he decreed, "will forever remain the mountain of the people of Maine."

The results of Baxter's foresight and stubbornness are impressive. The park roads are still dirt, with speed limits of 10 to 20 miles per hour; there are no hotels, no hot-dog or souvenir stands; outsized trailers, motorcycles, motorboats and other recreational paraphernalia, including firearms, are not allowed. Camping is restricted to designated, closely managed campsites; advance reservations are accepted from Maine residents from January 1 to March 31, and from out-of-staters after that. No more than a thousand people can spend the night at one time within the 200,000-acre area; when the campsites are full you are politely turned away at the gate. Baxter, as he said, wanted people to return to nature simply—or go somewhere else.

The park's loveliest enclave—the Wassataquoik Valley at the heart of the preserve—lies several miles to the north of Baxter Peak. Here, cradled in lower but no less beautiful mountains, are remote ponds and lakes and wild, tumbling streams. Here, too, is an abundance of moose, deer, beaver, bear and other wildlife, enjoying total protection under the terms of Baxter's will. At Wassataquoik Lake, a five-hour hike up from the trailhead at Roaring Brook campground, there is a little island surrounded by clear blue water; the only sound is that of Green Falls cascading down the mountainside a half mile up the lake.

It is hard to believe that even a sanctum such as this was invaded by the loggers, but in the old days, long before Percival Baxter's time, scenic spots were a dime a dozen. The hillsides were denuded of trees, and Wassataquoik Stream was blasted with dynamite to make passage for the logs. The last timber operations here stopped in the 1920s and the recuperative power of the forest has virtually restored the valley. But there are a few traces—a boulder still sharp-edged from dynamite, a tote road not quite overgrown.

The river drives in Maine, and the rest of New England, are all but gone today. It is cheaper to move the logs by big high-speed diesel trucks, and public pressure has demanded that the rivers be cleaned up and made available for fishing and canoeing once again. Great Northern, the state's biggest timberland owner, held its last drive of spruce and fir pulpwood down the Penobscot's West Branch in 1970. The Maine legislature has decreed that all river drives must end and that new pollution controls be in effect by 1976.

This is not to say that the lumber business is dead—not by a long shot. It is still the backbone of Maine's economy, a billion-dollar industry that produces everything from pulpwood for paper products (now two thirds of the total) to construction lumber, plywood, wooden ice-cream spoons and 100 million toothpicks a day. The whole northern half of the state, some 10.5 million acres of wild land, is almost entirely privately owned for its timber crop. It is, in effect, a gigantic tree farm—a garden, as some industry foresters call it—where specimens that reach a certain age are systematically harvested by chain saw, mechanical skidder and "slashmobile" to make room for more. Happily, very little of the garden is farmed at once—it takes 30 to 50 years or more to grow a crop, depending on the species and the use to which it is put—and the rest is open for hikers, hunters, fishermen and canoeists to enjoy.

One way to see Maine's wild timberlands, and about the only way to grasp the size of them, is to fly over the territory, as I did one September day in a float plane out of Shin Pond, near Patten, Maine. It is still a nearly continuous forest, a great dark-green blanket stretching as far as the eye can see, handsomely embroidered in silver by glinting lakes and streams.

Ray Porter, the pilot of the plane and a woodsman who has been flying this country for years, pointed out a series of beaver dams that had stopped up a stream in a striking, ladder-like pattern. In a widening stretch of a river a little farther on, we could see two tiny black specks, with plumes of muddied water eddying out downstream behind them. "Moose," Ray explained. The most interesting patterns, however—ones that can be seen clearly only from the air—were those marking the slow, inexorable death-throes of the glacier-born ponds and smaller lakes. Most that we could see were still open, clear water. Others exhibited the first telltale scums of algae, often in lovely bloomlike blobs; in still others the bottoms had filled with enough decayed matter for a fringe of water plants to find a footing and begin its encirclement from shore. In ponds at a more advanced stage, the water had been reduced to a sinuous trickle down the center, and a few were solid yellowish mats, with the vanguard of the forest starting its march inward to reoccupy them for good.

The largest of the lakes northwest of Baxter—Telos, Chamberlain, Eagle, Churchill—form the upper end of the Allagash Wilderness Waterway, a 92-mile corridor reaching from near a corner of the park to where the Allagash River, flowing north, joins the St. John. The lower end of the Allagash would have become a lake too, if a proposed power dam at Rankin Rapids had gone through. After much wrangling among federal and state officials and timberland owners, Maine passed a bond issue, matched by federal funds, and the Allagash is now a state preserve, the first in the Northeast to be incorporated into the National Wild and Scenic Rivers System. Logging is prohibited for 400 to 800 feet back from either bank, and for a mile back only state-approved selective cutting is permitted. Seven or eight thousand canoeists, the waterway's superintendent told me, now make the trip every year.

The Allagash is still a beautiful river journey, particularly if you manage to avoid the summer crowds. But the first half is largely lake paddling, and along the overall route there is only one earnest stretch of white water, the nine-mile run called Chase Rapids. For a combination of length, loneliness and unspoiled beauty, spiced by rapids most of the way, a better proposition is its sister stream, the upper St. John.

This bizarre-looking "totem pole" in northern Maine is, in fact, the work of hungry beavers. As winter draws on and their stored food becomes scarcer, the beavers sometimes venture out through the snow and nibble the bark of tree trunks. In this case, they probably felled a tree when it stood three or four feet deep in snow, after a number of earlier attempts marked by the notches at lower snow levels.

Famous though the Allagash may be, not many people have even heard of the upper St. John. The entire St. John River is 450 miles long, which makes it 50 miles longer than the Connecticut and, next to the St. Lawrence, the longest river in the Northeast. It starts in the remotest corner of Maine, loops north and east to form part of the border with Canada, then sweeps south and east through New Brunswick to the Bay of Fundy at St. John. It was named by the Frenchman Samuel de Champlain, who was nosing along the coast in 1604 and happened to sail into its mouth on the feast day of Saint John the Baptist. For the upper St. John, which Champlain never saw, the original Indian name, Wollastook, is far more descriptive: it means beautiful river.

There are several good reasons why you should not attempt the upper St. John. It is difficult to get to, unless you fly in with your boats and gear, and the trip in any case requires planning and arrangements in advance. Several of the rapids call for above-average canoeing skill. And unlike those of the Allagash, the St. John's ponds hold precious little liquid, and there are no dams to regulate the flow. As a result the runoff in the upper watershed is rapid, and so the surest time to try the St. John is in spring just after the ice goes out—if you are prepared to handle the torrent. In early June and July the black flies, mosquitoes and

A beaver lodge in Baxter State Park stands ready for winter, with new twigs added on top and a food pile of leafy branches at right.

midges (*no-see-ums,* as the Indians aptly called them) should be considered; they can be terrible. In late summer the flies are largely gone, but so is the water, and you may have to drag your canoe, wading over the rocks, or pack up and go home. As they say in Maine, you have to pick your time to run the St. John. And probably the best time, all things considered, is autumn.

The four of us who were to make the trip met for dinner on Sunday, October 1, in the little village of Kingfield, Maine. Our leader was Gardner Defoe, an outdoorsman, teacher and licensed Maine guide who runs a ski dorm in Kingfield in winter and canoe camps for boys and girls in summer at Birch Island, farther north. I was to paddle bow in Gardner's canoe. Stern man for the other canoe, and official cook, was Dutch Demshar, a lean Pennsylvanian and ex-Navy pilot who had adopted the Maine woods. His bowman was Dix Leeson, who had cheerfully cut classes at the Yale Forestry School and brought along his own favorite paddle at the very mention of the trip.

There was frost on the ground before dawn Monday morning when we lashed our two aluminum canoes on top of one car, stowed our packbaskets and food in another, and took off. With us were Barry and Mike, two friends who were to drive the cars home and meet us a week later at the northern terminus of the trip. Baker Lake, we had decided after several phone calls to woodsmen and rangers up the line, was the highest point on the river where we stood an even chance of enough water at this time of year. The trip to Baker was circuitous, involving a four-hour drive up into Canada and back into Maine again on an International Paper Company logging road, permit and fee required in advance. On private logging roads, we were reminded at the gate, the logging trucks have the right of way, a rule we had no urge to dispute. ("If you think those guys are driving fast," said Dutch, "you should see the roads on Friday night, when they're headed home for Canada with a week's pay in their pants.")

By the time we reached Baker Lake the sun had warmed up to a still, lovely Indian-summer day. "How's the water down below?" Gardner asked Gordon Hardy, the forest ranger at the lake. "A little bony down to Southwest Branch," Hardy replied. "But you ought to get through with poles." We packed our gear in the canoes, stuck the long setting poles and extra paddles through the thwarts, waved goodbye to Barry and Mike, and pushed off down Baker Branch.

I had canoed some before, but I had never fully realized what a superlative way it is to travel, especially through wilderness. To the Indians it was a way of life: The canoes they built—out of spruce or cedar ribbing covered with birchbark and caulked with pine tar—took them fishing and hunting, upstream into the woods and downstream to the sea with the cycle of the seasons. Their craft were marvelous inventions: fast and silent enough to pursue a moose, light enough to carry on long portages, shallow enough to coast over rocks and gravel beds, sturdy enough to hold families and possessions. The modern canoe retains most of these virtues, and being made of aluminum, molded glass fiber or canvas-on-wood, allows more leeway for blunders than the fragile birch. Once you get into the swing of it, the simple act of propelling one of these eager craft can be a joy. After a day or two I had abandoned my awkward, all-arms stroke in favor of the less tiring Indian paddle Gardner taught me, holding the lower arm straight, using the whole upper body and rocking forward in a pushing, sculling motion. It's a nice sound in the river silence, the paddle shafts softly thumping the side of the canoe in cadence at the end of every stroke.

The spruces and firs on the St. John are tall and soldierly and black against the sky, and the fragrance of balsam and cedar drifted out to us from shore. From time to time we spotted the whorling, pagoda-like branches of a solitary white pine rising above the blanket of trees. Imperfect specimens passed over by the loggers, left to grow old and scraggly, these trees were still noble reminders of the scale of the earlier wilds. They pay a price for their height and solitude: more than one we saw had been slashed or toppled by lightning strikes.

As we paddled the upper river, we welcomed each new brook and streamlet along the way, for the St. John was indeed "bony" in spots from lack of recent rain, and we needed all the water we could get. When the canoes began to scrape over rocks and gravel bars, the setting poles would come out. In the hands of an experienced riverman this slender, supple length of spruce or ash, 10 or 12 feet long and capped at its lower end with steel, is an indispensable tool for poking in among the rocks to slow a canoe, move it into a deeper channel, nurse it over submerged boulders and, when all else fails, to back up and try again.

One of the pleasant things about river canoeing is that most of the time you can go at your own, or the river's, speed, looking at the scenery, bird-watching, daydreaming, examining the water or the sky or whatever else comes to mind. There is no need to race, unless you want to, and before the light begins to fail you can usually find a proper campsite along the shore. Among the prettiest sites we discovered was

A mature bull moose, his large, well-shaped antlers reflecting his prime physical condition, trots across a field in Baxter State Park.

one located on a rise where the Southwest Branch joins the Baker Branch, swelling the river to a respectable width and depth. Although the water everywhere along the upper St. John is cold, clear and pure enough to cup to your lips, Gardner knew of a special spring off through the woods, and before making camp we welcomed the excuse to stretch our legs. Along the path in a clearing we came upon wild onions, rose bushes with the flowers gone and bright-red rose hips in their places, and a few withered raspberries still clinging to bushes that must have provided some bear with a tasty meal. On fallen and decaying tree trunks there had sprung up little colonies of plants: tiny mosses of every description, topped here and there by the red-capped lichen called British Soldiers, forming fantastic miniature forests within the larger one. At the spring we filled the water bag with cold, sweet liquid and turned back through the woods to the shore.

Making camp after a long day on the river has special satisfactions of its own. We soon fell into a routine. Dix and I would find a level place and set up the two "pop" tents, collapsible wigwams of light canvas with glass-fiber ribs. The others would gather firewood, and saw and split enough of it for supper and breakfast. Then with a flourish Dutch would unpack his prize utensil, a gigantic frying pan, get a pot of coffee going and start biscuits in the reflector oven. In the early October evenings there was a lot of stamping of feet, rubbing of hands, and trying to keep wet socks and shoes from falling into the fire.

Dutch's dinners were surprisingly good, from stews to apple pie, accompanied by a thoughtfully chosen bottle of Riesling or Beaujolais (you can carry a lot more in a canoe than in a knapsack). Breakfasts consisted mainly of eggs, Canadian bacon and more hot coffee, and tasted better than any I have ever had. The lunches we made and packed along, on the other hand, were as unchanging as the hills: peanut butter and jelly, ham on rye and cherry Kool-Aid, with Life Savers for dessert. "Quick energy," Dutch said.

Out on the river in the mornings, as the early mists burned off, the water would come alive. Sandpipers and yellowlegs flitted and hopped about close to shore. Around almost every bend we came on flights of larger birds resting on their flyway routes; black ducks, goldeneyes and mergansers rose flapping from the surface and flew swiftly over or around us, resuming their journeys south. Far down the glassy stillness of a deadwater, a great blue heron would spread its six-foot wings and lazily disappear. It became a game to see how close we could get to flocks of aristocratic buff-and-black Canada geese before they honked their displeasure and flew off, forming southbound chevrons in the sky. Perched in a riverbank aspen, or popple as Mainers call it, was a pair of horned owls; as we approached a rocky promontory we surprised three handsome, husky white-tailed deer, who eyed us briefly, clattered up the rocks and were gone.

Along the shoreline, particularly on the outsides of bends, we had noticed signs of the growing river's power: tree trunks scarred and whitened 6 to 10 feet above water level by the annual spring breakup and downstream stampede of the ice. Just above Nine Mile Brook we saw evidence of how strong the river could be. A logging-road bridge there, once solidly built of steel, had been reduced to its rubble-filled wooden piers in what must have been a champion "ice-out" a few years before. One giant steel truss still lay on the bank a hundred yards below; the other had been carried a full three miles downstream and deposited on the opposite shore.

In her book, *Nine Mile Bridge,* Helen Hamlin, who spent a year or two in the cabin at Nine Mile as a game warden's wife, described the spectacle of an ice-out here, which she started to watch from an earlier bridge. "There wasn't much to see at first, but soon the faint roar became thunderous and angry. A ten-foot wall of tumbling, crackling, fast-moving ice rolled around the upper bend of the river, sweeping everything before it, gathering momentum and throwing two-ton floes on the high banks. It uprooted trees and crumbled the fettered ice sheet in its path. The dammed water behind pushed relentlessly, increasing in force and power, and coming closer and closer.

"It was greenish white, and some of the packed, moving ice reflected the sun's rays in rainbow colors. As far as we could see this massed ice plateau was surging—rising as high as the riverbanks. The ten-foot wall grew to twelve feet, then to fifteen, and thirty feet from the bridge it angrily greeted us with an ear-splitting screech. Curly touched my arm and I turned away. The spell was broken and we started to run."

That year the bridge held; Curly, the warden, even managed to save a doe trapped among the ice floes. Mrs. Hamlin won a pair of pinking shears from her husband for guessing the ice-out to the day, April 27; but she noted with regret that it left a 20-foot-high pile of floes on their front lawn, cutting off the view. The shuttered cabin was still standing when we pulled ashore there to eat our sandwiches, and the view from the lawn was peaceful and iceless in an October midday sun.

Below Nine Mile, the river left its dark spruce-fir corridor to broaden and meander around Seven Islands, an unexpectedly pastoral place of waving yellow meadow grasses and upward-arching elms. The scene, with its wide vistas and ancient trees, reminded me of nothing so much as a stately, well-kept English farm. Back in the river-driving days half a century or more ago, the islands were in fact a farm, the site of a thriving community that supplied the loggers upwoods with potatoes, poultry, milk and beef. But on the treeless plain where we camped, the buildings had long since crumbled and disappeared; the only relic we could find was quite a different one: the twisted airframe of a light plane that had crashed and burned trying to land on the uneven meadow some years earlier. On the islands we saw more flights of water birds, and I shared the river for my morning wash-up with a large and rather shy cow moose. There is supposed to be fine trout fishing in many of the brook- or spring-fed bogans, inlets branching off the main stream, but we had missed the end of the fishing season by a few days.

Below Seven Islands the forest closed in again, although the river itself was now a stately hundred yards or more across, moving with deliberate speed as it approached the major white waters in its course. We could hear Big Black Rapids long before we saw them—the low, rushing roar an immense volume of water makes as it gains momentum on a long downhill slide, dropping and pounding over thousands of tons of rocks. Around a bend we came upon the rapids: big and black, with jagged rows of white rips and spume tossing from the churning waves canoeists call haystacks and rooster tails. The river, until now alternately placid and mildly tricky, was beginning to show its teeth.

Although he had run Big Black many times before, Gardner pulled ashore like any sensible riverman to scout it first on foot; the best channels vary from year to year and from season to season with the level of the water, and a heavily loaded canoe allows even less than the usual margin for error. We clambered along the rocky bank for half a mile downstream, stopping to check a rock here, a possible place to squeeze through there, an ugly rip somewhere else. Gardner seemed lost in thought, massaging his chin and making occasional fish-tailing motions in the air with his hands as he maneuvered an imaginary canoe on a new route. Looking at the river, I recalled reading somewhere that water moving at only 10 miles an hour could press a capsized canoe against a rock with a force of seven tons, bending it in half like a paper clip. I resolved to disembark, if I had to, on the upstream side of my craft. Someone said a couple of men had drowned here a few years back. I reviewed

An early-morning mist shrouds a channel of the St. John River in the Seven Islands area of northern Maine. At this point the river has flattened out, widened and slowed down, dropping sediment that has built up to form numerous small islands. The river here wends a lazy course between stands of white pines, eerie skeletons of dead spruces and the gold of turning aspens on the slope of a hill.

the procedures for swimming downstream with feet forward to fend off rocks. I also considered the ultimate eventuality: without canoes and food it could be a very long, cold walk home.

At length our leader seemed to have a plan, and we trooped back to the canoes, took another notch in our lifebelts and shoved off. "See that big pointy rock over there?" Gardner shouted to me over the noise of the water from the stern. "Let's try her just to the right of that. Steady . . . steady, draw left a little. Left, dammit, not right! *Now* draw right! Hard. Harder! Now draw left, *hard!*" We were threading our way amongst some frightening chunks of granite, following a curling tongue of water that hissed and snaked and boiled through a stony obstacle course. Now, to keep control, we needed speed greater than that of the fast-moving water. "Paddle," yelled Gardner. "Paddle like hell!" I dug the blade in until my shoulders threatened to come unglued. When we emerged from the mile-and-a-half maelstrom of Big Black, I was sure we were exceeding the posted speed limit for the state. "Pretty good sleigh ride, eh?" Gardner grinned. It was. Next year, Niagara. We pulled in that afternoon near an abandoned logging camp; a cold rain was driving down. After dinner I crawled into the tent and slept for 10 hours without opening an eye.

The rain continued the next day and the river rose, sweeping us along. On the shores low-bush blueberries were turning crimson, and small birches, like tiny skyrockets, detonated their yellow leaves against the dark woods. Now and then the rain and mist lifted: the high hillsides that suddenly rose above us were carpeted with a tapestry of colors, of brilliances and gradations I had never seen before—the lime-green and gold of aspen and birch, the smoky purple of ash, the oranges and reds of maples. Every lovely color was held and rendered doubly brilliant by the dark matrix of the evergreens.

I think if I had to show someone New England only at one instant, in one time and place, it would have to be this: from a canoe suspended on a silver river, surrounded by the great, silent autumnal explosion of the trees. On the hills the evergreens stand unchanging, ancient, absolute; their tough, resinous needles, evolved millions of years ago, are ready to take winter at its worst. Scattered in abstract patterns through their ranks, the deciduous trees, which have developed more sophisticated systems, have finished their spurt of summer growth and are preparing to drop their broad, life-giving leaves and hibernate. Signaled by the shortening daylight, a corky barrier has formed in each leaf stalk to strangle off the sap; the green of chlorophyll fades, unmasking the underlying chemicals that "turn" the leaves yellow or, combined with others like sugars and tannins, produce the glorious golds, oranges, reds and purples that determine the regimental colors of the species. This year a gradual onset of cool weather, warmed by sunny days, has allowed the dying leaves to progress through the full range of their display, uninterrupted by killing frosts. The result is a New England autumn, unmatched anywhere else on earth. No matter how many times you have seen it before, it is an astonishing sight: the army of trees drawn up along the hills, bearing its brief flags of fall.

Near the Canadian border the St. John tumbles through its last major white water, a two-mile stretch called Big Rapids, then emerges from the forest. The flaming hillsides yield to tranquil farms. The wilderness is gone. The valley narrows momentarily near the settlement of Dickey; this is the site of the proposed power dam the Allagash wouldn't have. If ever built, it would wipe out the last 50 miles of river we have just run, drowning the trees and rapids, burying Seven Islands, driving the animals back into the hills. It would ruin, once and for all, the finest wild river left in the Northeast.

By the time we reached the junction with the Allagash, the river had risen a full foot from the rains; riding the swell with a following wind, we flew over Golden and Rankin rapids without even seeing the rocks. At the village of St. Francis, a little border settlement nestled in the notch in the top of Maine, we took the canoes out for the last time and walked up the bank to Pelletier's general store to get hot coffee and a Sunday paper and wait for the cars. "How was the trip, boys?" asked the proprietor. "Beautiful," someone said. It was. Eight days on the beautiful river.

Down on the landing the wind rose, whirling the first flakes of an early snowstorm in our faces. We lashed the canoes on top of the car and headed home.

Katahdin's Autumn

PHOTOGRAPHS BY DAN BUDNIK

The theatrically brilliant autumns so famous in New England, brought on by sharp, chilly nights and gentle, sun-warmed days, are nowhere more stunning than in Maine's Baxter State Park. There Mount Katahdin presides grandly over a landscape that in late September and early October fairly catches fire with the changing colors of deciduous trees. The spectrum ranges from the delicate yellow of birch to the velvety purple of American ash, including almost every shade between; many of the most luminous oranges, reds and yellows appear simultaneously on the sugar maple, the region's most spectacular tree.

Photographer Dan Budnik was so intoxicated by Baxter's riot of color that he spent a week there attempting to catch the full impact on film. One maple tree in particular *(pages 176-177)* so attracted him that he flew over it nearly 20 times in a light plane. "I felt like a hawk soaring back and forth until I began to lose all sense of where I was," he says. "It was like viewing a living piece of sculpture that kept changing as the light caught it in different ways."

During Budnik's explorations high winds blew many of the leaves off the trees before their brilliance had a chance to fade, providing an unexpected impression: "One evening, walking towards Roaring Brook, I was mesmerized by the leaves on the ground in the weak afterglow of sunset. It was almost dark, yet the leaves seemed still aflame, alive, their colors playing off against each other and against the warm browns of the earth. It took me twice as long to reach camp as I had planned."

The photographer was struck by other signals of dying summer and approaching cold: ferns becoming a rich bronze, and reeds and sedges turning tawny yellows; the brassy glow of an autumn sun rising on a pond misted in the chill morning air; lakes, usually mirror-like, whipped into white-fringed waves beneath scudding storm clouds.

Everywhere Budnik's eye kept coming back to the raw beauty of the mountain itself, rearing high above the forest. Late one afternoon he flew over the sweep of its three main peaks: "The sun kept getting lower and lower, the light more golden. Slowly the long shadow of the mountain inched eastward over Katahdin Lake, reaching toward Canada and the sea. The bare, rocky rubble cradled in the summits was the last place to hold the dying rays."

FLAMING MOUNTAIN ASH AGAINST GREEN SPRUCE

WIND-RUFFLED REEDS AND WATER AT TRACY POND

DROWNED SPRUCES AT A BEAVER POND

RED-MAPLE LEAVES ON BRONZE FERNS

A LEAF IN A CLEAR POND

SUNRISE OVER LOWER TOGUE POND

WHITECAPS ON UPPER TOGUE POND

BIRCHES AND PINES ON A CLIFF OF JAGGED RHYOLITE

A BLAZING SUGAR MAPLE FROM THE AIR

SUNSET ON MOUNT KATAHDIN

Bibliography

**Also available in paperback.*
†Available only in paperback.

A.M.C. New England Canoeing Guide. Appalachian Mountain Club, 1971.

A.M.C. White Mountain Guide. Appalachian Mountain Club, 1972.

Amos, William H., *The Life of the Pond.* McGraw-Hill Book Company, 1967.

*Baxter, Constance, and others, *Greatest Mountain: Katahdin's Wilderness.* Scrimshaw Press, 1972.

*Bearse, Ray (ed.), *Massachusetts: A Guide to the Pilgrim State.* Houghton Mifflin Company, 1971.

*Bearse, Ray (ed.), *Vermont: A Guide to the Green Mountain State.* Houghton Mifflin Company, 1968.

Bent, Arthur Cleveland, *Life Histories of North American Birds of Prey.* 2 vols. Dover Publications, 1961.

Bent, Arthur Cleveland, *Life Histories of North American Diving Birds.* Dover Publications, 1963.

†Billings, Marland P., Katharine Fowler-Billings, Carleton A. and Randolph W. Chapman, and Richard P. Goldthwait, *The Geology of the Mt. Washington Quadrangle, New Hampshire.* The New Hampshire Planning and Development Commission, 1972.

†Bliss, L. C., *Alpine Zone of the Presidential Range.* Edmonton, Canada, 1963.

†*A Brief Guide to the Natural History of the White Mountains.* Audubon Society of New Hampshire, 1967.

Bullaty, Sonja, Angelo Lomeo, and Noel Perrin, *Vermont in All Weathers.* The Viking Press, 1973.

Burt, F. Allen, *The Story of Mount Washington.* University Press of New England, 1968.

Burt, William H., and Richard P. Grossenheider, *A Field Guide to the Mammals.* Houghton Mifflin Company, 1964.

Chapman, Frank M., *Handbook of Birds of Eastern North America.* Dover Publications, 1966.

Conant, Roger, *A Field Guide to Reptiles and Amphibians.* Houghton Mifflin Company, 1958.

Dietz, Lew, *The Allagash.* Holt, Rinehart and Winston, 1968.

*Doan, Daniel, *Fifty Hikes.* New Hampshire Publishing Company, 1972.

†Douglas, William O., *My Wilderness: East to Katahdin.* Pyramid Books, 1968.

Eckstorm, Fannie Hardy, *The Penobscot Man.* New Hampshire Publishing Company, 1972.

Fisher, Dorothy Canfield, *Vermont Tradition.* Little, Brown and Company, 1953.

Gleason, Henry A., and Arthur Cronquist, *The Natural Geography of Plants.* Columbia University Press, 1964.

†*Guide Book of the Long Trail.* Green Mountain Club, 1972.

†*Guide to the Appalachian Trail in Maine.* The Appalachian Trail Conference, 1969.

†*Guide to the Appalachian Trail in Massachusetts and Connecticut.* The Appalachian Trail Conference, 1972.

†*Guide to the Appalachian Trail in New Hampshire and Vermont.* The Appalachian Trail Conference, 1968.

Hamlin, Helen, *Nine Mile Bridge.* W.W. Norton and Company, 1945.

*Hard, Walter, Jr. (ed.), *Green Mountain Treasury: A Vermont Life Sampler.* Harper and Brothers, 1961.

Hill, Ralph N., M. Hoyt, and W. R. Hard Jr., *Vermont: A Special World.* Vermont Magazine, 1969.

Holbrook, Stewart H., *Holy Old Mackinaw.* Macmillan Company, 1956.

Holbrook, Stewart H., *Yankee Loggers.* International Paper Company, 1961.

Hubbard, Lucius L., *Woods and Lakes of Maine.* New Hampshire Publishing Company, 1971.

*Jorgensen, Neil, *A Guide to New England's Landscape.* Barre Publishing Company, 1971.

Lowenthal, David, *George Perkins Marsh: Versatile Vermonter.* Columbia University Press, 1958.

Marsh, George Perkins, *Man and Nature.* Harvard University Press, 1965.

†*Mount Washington.* Mount Washington Observatory, 1962.

Mountain Flowers of New England. Appalachian Mountain Club, 1964.

Niering, William, *The Life of the Marsh.* McGraw-Hill Book Company, 1966.

Oakes, William, *Scenery of the White Mountains.* New Hampshire Publishing Company, 1970.

Peattie, Donald Culross, *A Natural History of Trees of Eastern and Central North America.* Houghton Mifflin Company, 1963.

Peterson, Roger Tory, and Margaret McKenny, *A Field Guide to Wildflowers.* Houghton Mifflin Company, 1968.

Pike, Robert E., *Tall Trees, Tough Men.* W.W. Norton and Company, 1967.

Rood, Ronald, and others, *Vermont Life Book of Nature.* The Stephen Greene Press, 1967.

Smith, Chard Powers, *The Housatonic.* Rinehart and Company, 1946.

Smith, David C., *A History of Lumbering in Maine 1861-1960.* University of Maine Press, 1972.

Springer, John S., *Forest Life and Forest Trees.* New Hampshire Publishing Company, 1971.

Sutton, Ann and Myron, *The Appalachian Trail: Wilderness on the Doorstep.* J.B. Lippincott Company, 1967.

Thomson, Betty Flanders, *The Changing Face of New England.* Macmillan and Company, 1968.

*Thoreau, Henry David, *The Maine Woods.* Thomas Y. Crowell Company, 1966.

†*Winter on Mount Washington.* Mount Washington Observatory, 1970.

Zwinger, Ann H., and Beatrice E. Willard, *Land above the Trees.* Harper and Row, 1972.

Acknowledgments

The author and editors of this book are particularly indebted to Hubert W. Vogelmann, Professor of Ecology, University of Vermont, Burlington; Malcolm W. Coulter, Associate Director for Wildlife, School of Forest Resources, University of Maine, Orono; John Nutter, Director of Education, Appalachian Mountain Club, Pinkham Notch, New Hampshire; and Sidney S. Horenstein, Department of Invertebrate Paleontology, The American Museum of Natural History, New York City. They also wish to thank the following individuals and institutions. In Maine: Constantine Albans, Penobscot Heritage Museum, Bangor; Irvin C. Caverly Jr., Baxter State Park, Millinocket; John N. Cole, *Maine Times,* Topsham; Alyce M. Connor, Bangor Public Library; Gardner Defoe, Kingfield; Don Fletcher and Don McKay, Diamond International Co., Old Town; John Maines, Brewer; Maynard F. Marsh, Department of Inland Fisheries and Game, Augusta; John McKee, Brunswick; Lore A. and Katherine Rogers, The Lumberman's Museum, Patten; James B. Vickery, Brewer. At the University of Maine, Orono: Eva K. Dimond, Raymond H. Fogler Library; Edward Ives, Northeast Archives of Folklore and Oral History; David C. Smith, Professor of History. In New Hampshire: Paul Bofinger, Society for the Protection of New Hampshire Forests, Concord; Tom Deans, Appalachian Mountain Club, Pinkham Notch; Paul Doherty, Hilbert Siegler and Helenette Silver, New Hampshire Fish and Game Department, Concord; John Howe, Mount Washington Observatory; Ken Norcott, Brown Paper Co., Berlin; Tudor Richards, Audubon Society of New Hampshire, Concord; Frederic Steele, White Mountain School, Littleton; Ned Therrien, U.S. Forest Service, White Mountain National Forest, Concord. In Vermont: Walter R. Hard Jr., *Vermont Life,* Montpelier; Frederick H. Mold, Howard B. Reed and William C. Christiansen, Fairbanks Museum of Natural Science, St. Johnsbury; Frank Morrell, U.S. Forest Service, Green Mountain National Forest, Rutland; John D. Randolph, *The Vermont Sportsman,* Bennington; Robert M. Smith, Vermont Fish and Game Department, Lunenberg. In Massachusetts: C. Francis Belcher and Malcolm C. Choate, Appalachian Mountain Club, Boston; Perry Hagenstein and Robert August, New England Natural Resources Center, Boston. In Connecticut: Frank Calhoun, Cornwall; Paul H. Chamberlain Jr., Cornwall Bridge; David M. Smith, Yale Forestry School, New Haven. In New York City: Joseph A. Davis, New York Zoological Society; Davis Finch, National Audubon Society; James E. Miller, Polytechnic Institute of New York; Larry G. Pardue, New York Botanical Garden. Also: Peter H. Dunning, Appalachian Trail Conference, Harpers Ferry, West Virginia; Richard Saltonstall, Great Falls, Virginia; Jean Stephenson, Washington, D.C. Special thanks are due those who participated in the rescue mission for photographer Dean Brown: Verland Ohlson, District Ranger, and Homer Emery, Bill Gibson, Allan Green, Wendell Lees, Arlyn Perkey, Robert Smith Jr., Don West and Rick Young, all of the U.S. Forest Service, Conway, New Hampshire; David Beyerle and Henry Mock, New Hampshire Fish and Game, Madison; Raymond Nitz, Carroll County Sheriff's Office, Ossipee; Tom Barringer and Greg Betts, Appalachian Mountain Club, Pinkham Notch.

Picture Credits

Sources for the pictures in this book are shown below. Credits for pictures from left to right are separated by semicolons; from top to bottom they are separated by dashes.

Cover–Dan Budnik. End papers 2, 3 –Fred B. Stimson III. End paper 4, page 1–Dean Brown. 2, 3–Dan Budnik. 4, 5–Dean Brown. 6, 7–Dan Budnik. 8, 9–Dean Brown. 10, 11–Sonja Bullaty and Angelo Lomeo. 12, 13 –Fred B. Stimson III. 18, 19–Maps produced by Hunting Surveys Limited. 22–Fred B. Stimson III. 26, 27–Fred B. Stimson III. 33 through 43–Sonja Bullaty and Angelo Lomeo. 47, 48, 49 –Henry Groskinsky. 52–Ogden Tanner. 56, 57–Sonja Bullaty and Angelo Lomeo. 60 through 71–Robert Walch. 74, 75–Ned Therrien. 79–Dean Brown. 82–Jeff Foott from Bruce Coleman, Inc. 84, 85–Dean Brown. 89 through 103 –Dean Brown. 106–Clyde H. Smith. 110, 111–Bob Candy. 113–Paul Doherty. 114–Barbara K. Deans from National Audubon Society. 119–David G. Allen. 120, 121–Fred B. Stimson III. 122, 123–John E. Swedberg; Carl E. Scott and John E. Swedberg. 124, 125 –Fred B. Stimson III; John E. Swedberg. 129–Clyde H. Smith. 132, 133 –Dean Brown. 137–Courtesy The Lumberman's Museum, Patten, Maine. 138–Courtesy Northeast Archives of Folklore and Oral History, University of Maine at Orono. 139–Courtesy Special Collections, Fogler Library, University of Maine at Orono. 140, 141 –George H. Hallowell, courtesy The Lumberman's Museum, Patten, Maine; Courtesy The Lumberman's Museum, Patten, Maine. 142, 143–Courtesy The Lumberman's Museum, Patten, Maine; George H. Hallowell, courtesy The Lumberman's Museum, Patten, Maine. 144, 145–George H. Hallowell from The Myron H. Avery Collection, Maine State Library. 150, 151–Kit Luce. 154, 155–Dan Budnik. 158–Dan Budnik. 162, 163–Clyde H. Smith. 167 through 179–Dan Budnik.

Index

Numerals in italics indicate a photograph or drawing of the subject mentioned.

A
Agassiz, Louis, 50
Alder, 128, 131
Allagash Wilderness Waterway, 154, 165
Androscoggin River, 126, 128
Animals, threatened and thriving species, 108-110, 112, 114-117
Appalachian Mountain Club, 76, 82-83, 84; hostelries, 82-83, 87
Appalachian Trail, 20, 60; highest section, 76-82; in Maine, 146, 147, 149; tourism, 83, 86
Arnold, Benedict, 147
Aroostook County, 147
Ash, 164; American, 166; mountain *(Pyrus americana), 167*
Aspen, *162-163,* 164
Aster, 26; mountain *(Aster acuminatus), 92-93*
Autumn, 166, *167-179;* tree changes, 164-165, 166
Avens, mountain *(Geum peckii), 8-9,* 78, *79*
Avery Peak, 147
Azalea, alpine *(Loiseleuria procumbens),* 77, 78, *79*
Aziscohos Lake, 147

B
Baker Lake, 156
Bangor, Me., 136, 148
Bangor Tigers, 136, *137-145,* 148
Barnum, P. T., quoted, 84
Barren-Chairback Range, 147
Baxter, Percival Proctor, 29; and Baxter State Park, 150, 152
Baxter Peak, 150
Baxter State Park, *cover, 6-7,* 150, *155, 158;* autumn, 166, *167-179;* creation, 150, 152
Bear, black, 110, 127, 152, 159
Beaver, 80, 109, 152, 153; gnawed tree, *154;* lodge, *155;* pond, 56, *169*
Beech(es), 21, 23, 26, 31, 32, *34,* 53, 56, 58, 60, 126; American, 132
Berkshires, *12-13,* 44, 50
Big Black Rapids, 161, 164
Bigelow Mountain, 146
Bigelow Range, 147
Billings Falls, *150-151*
Birch(es), 21, 32, *34, 42-43, 56-57, 70,* 126, 164, 166, *175;* cherry, 23; gray, 21; paper or white *(Betula papyrifera), 6-7,* 26, 56, 78, *92-93,* 132; yellow, 58, 60, 70, 132
Birds: of Alpine-Arctic zone, 80, 82; defenses, 118; and forests, 55; nesting, 118, *119-125;* number, 118; water, 159-160
Blackbird, red-winged, 131
Bladderwort *(Utricularia vulgaris), 64,* 65, 67
Bluet, alpine *(Houstonia caerulea),* 78, *79*
Bog laurel *(Kalmia polifolia), 64-65, 67*
Bogs, 60, *62-63, 66, 70-71;* decay inhibition, 64, 68; environment in, 60, 63-64; forest, *61, 66,* 67-70; formation, 60, 63-64; life forms, *60-71,* 128; mat, *62-63,* 64, 65; thicket, *66,* 67
Bonney Brook, 23
Bromley Mountain, 45
Bunchberry *(Cornus canadensis), 92-93*

C
Calhoun, J. F., 30
Calla, wild *(Calla palustris), 67*
Caribou, 109
Cathedral Pines, *end paper 2-end paper 3,* 26, 30-31; preservation, 30
Cedar, 157; red, 26
Champlain, Samuel de, 154
Chickadee, black-capped, 21
Chimney Pond, 150
Chipmunk, *122-123*
Chokeberry, purple, 67
Cinquefoil, dwarf *(Potentilla robbinsiana),* 78, *79*
Clintonia. *See* Lily.
Club moss, 77
Conservation, 29-30, 54-55, 59, 84, 128, 150, 151, 152-153, 154; federal government, 77; threatened and thriving animals, 108-110, 112, 114-117; and tourism, 86. *See also* Forests
Coot, 107
Coulter, Malcolm, 112
Coyote, eastern, 112, *113,* 114-116, 127, 131
Crawford, Abel and Ethan Allen, 84
Crawford Notch, 85

D
Damselfly *(Enallagma cyathigerum), 67*
Dandelion, 26
Dead River Valley, 147
Dean Ravine, *26-27*
Deer (white-tailed), 55, 56, 80, 109, 115, 128, 132, 152, 160
Deer Brook, *4-5*
Deerfield River, East Branch, *52,* 55, 56
Defoe, Gardner, 156; trip with, 156-157, 159-161, 164-165
Demshar, Dutch, 156
Diapensia, 77
Dogberry, 128
Douglas, William O., 150
Dragonfly *(Leucorhinia hudsonica), 67*
Drumlin, 51
Duck(s), 107; black, 159; goldeneye, 159; merganser, 159
Dudleytown, Conn., 21, *22, 23*

E
Eelgrass, 64
Elephant Mountain, 146
Ellis River, *96-97*
Emerson, Ralph Waldo, 54

F
Falsehellebore *(Veratrum viride), 42*
Fern(s), 30, 32, 87, 166, *170;* cinnamon *(Osmunda cinnamomea), 68-69;* evergreen, *42;* wood *(Dryopteris spinulosa),* 69, *92-93*
Field, Darby, 83
Fir(s), *40-41,* 53, 56, *70,* 86, *90,* 99, 126, 128, 131, 153, 157; adaptations, 78, 80; balsam *(Abies balsamea),* 26, 58, 60, 64, 67, *68-69,* 70, 131
Fisher *(Martes pennanti), 110-111,* 112, 114
Flagstaff Lake, 147
Fly, black, 65, 130, 154
Forests: bog, 67-70; changes since ice ages, 53; effects of, 54-55, 56, 58; Maine, 147, 153; management, 29-30, 54. *See also* Trees
Fox, red, 110
Franconia Range, 78
Frog(s), 131; eggs and tadpoles, *64,* 65; green *(Rana clamitans), 64,* 65; spring peeper, 65; wood *(Rana sylvatica), 62,* 63
Frost, Robert, quoted, 20

G
Geology: cirques, 95; drumlins, 51; erratics, 53, *92-93; Felsenmeer, 102-103;* landslides, 98, *100-101;* mountain formation, 50-51, 53, 98, 100; Taconic Revolution, 50. *See also* Glaciers; Rocks
Glaciers: and bogs, 60; erosion, 24, 51, 77, 90, 95, 149; erratics, 53
Goldthread, 69, 87
Goose, Canada, 160
Gramlich, Frank, 116
Grass(es), 26; meadow, 161; reeds, 166, 168; tundra, 53
Great Basin, 149
Great Gulf, 76-77, 86-87, *90, 100-101*
Grebe, 105
Green Falls, 152

Green Mountain National Forest, 55-56, 84
Green Mountains, 20, 32, *33-43*, 44; geologic history, 50-51, 53; and Vermont economy, 46
Grouse, 109, 115, 123, 128; ruffed *(Bonasa umbellus)*, 118, *119*
Gulf Hagas, 147, *150-151*

H
Hamlin, Helen, *Nine Mile Bridge*, 160
Hare, snowshoe, 56, 80
Hawk(s), 81; goshawk *(Accipiter gentilis)*, 118, *122-123*; marsh, 130; osprey *(Pandion haliaetus)*, 118, *123*, 130; rough-legged, 130
Hemlock, 21, 23, 31, *42-43*, 53, 131
Heron, great blue, 159
Hickory, 26, 53
Hobblebush, 87, 135
Holbrook, Stewart, quoted, 50
Housatonic State Forest, *26-27*
Howe, John, 72
Hunt Spur, 149

I
Ice age(s), 51, 53, 60, 77, 98, 102-103
Ice-out, 160
Indians, 24; Abnaki, 104, 148; canoes, 157; legend, 148; and logging, 148; and Mount Katahdin, 148; and Mount Washington, 83; Penobscot, 148

J
Jay, Canada, or whiskey jack *(Penisoreus canadensis)*, *114*, 116-117
Jay Peak, 45
Josselyn, John, quoted, 72
Junco, slate-colored, 80

K
Knife Edge, 149-150

L
Labrador tea, 64, 77, 78
Lake Umbagog, 30, 126, 128, *129*
Lake Willoughby, 51
Lapland rosebay, 77, 78, 88
Larch, 64; eastern, 131-132
Leatherleaf *(Chamaedaphne calyculata)*, 63, *64-65*
Leeson, Dix, 156
Lichen(s), 53, 70, 77, 81, *90, 100*; British Soldiers, 159; foliose *(Parmelia physodes)*, *70*; old man's beard, 86
Lily, clintonia, or bluebead *(Clintonia borealis)*, *68*, 69
Lily of the valley, wild, or Canada mayflower *(Maianthemum canadense)*, *68*, 69
Litchfield Hills, 21, 44
Logging, 25, 28-29, 53, 54, 84, 126-127, 128, 136, *137-145*, 148-149, 152-153, 154
Long Trail, 56, 60
Loon, common *(Gavia immer)*, 104-105, *106*, 107-108; protection of, 108
Lost Pond Bog, 60; life forms, *60-71*
Lynx: bobcat, 127; Canada, 127

M
Madison Gulf Trail, 86
Magalloway River, 126, 128
Mahoosuc Notch, 146
Maple(s), 21, 23, 26, 31, *34, 42-43*, 53, *124-125*, 126, 164; leaf, *33*; moosewood, 132; red *(Acer rubrum)*, 132, *170*; striped, 132; sugar *(Acer saccharum)*, *cover*, 16, 58, 60, 132, 166, *176-177*
Marsh, George Perkins, 29, 54; *Man and Nature*, 54-55
Marten, 127; Pennant's (fisher), 110; pine, 110
Matt, Toni, 86
Minerals: limonite, *150-151*; mica, 77, 88; quartz, 46, *48*, 81; serpentine, 51; tourmaline, 46, *48*. *See also* Rocks
Mitchell, Big Sebattis, 148-149
Mold, Fred, 128
Moose *(Alces alces)*, 109, 127, 132, 152, 153, *158*, 161
Moose River, 127, 128
Moosehead Lake, 147
Mooselookmeguntic Lake, 147
Mosquitoes, 65, 154
Moss, 60, *68-69*, 86, *90*, 159; club, 77; peat, 63; sphagnum, 60, *62-63*, 64
Moss plant *(Cassiope hypnoides)*, 78, *79*
Mount Adams, 72, 81
Mount Cleveland, 45
Mount Grant, 45
Mount Hor, 51
Mount Horrid, 46
Mount Jefferson, 72, 81
Mount Katahdin, *2-3*, 29, 147-148, 149, 166, *178-179*
Mount Lincoln, 45
Mount Madison, 72
Mount Mansfield, *10-11*, 36, *38*, 39, *40-41*, 45
Mount Pisgah, 51
Mount Roosevelt, 45
Mount Spaulding, 146
Mount Washington, *8-9*, 72, *74-75, 100, 102-103*; animals, 80, *82*; dangers hiking, 73; exploration and development, 83-84, 86; vegetation, *8-9*, 77, 78, *79*, 80, *90*; weather, *72-73*, 76
Mount Wilson, 45
Mountain holly, 67
Mountain laurel, 23
Mountain lion, eastern, 24, 108-109
Muir, John, 55
Mumford, Lewis, quoted, 55
Muskrat, 108, 128

N
National Wild and Scenic Rivers System, 154
Nelson Crag, *74-75*
Nesowadnehunk Falls, 148
Nesowadnehunk Stream, 149
North country, 126-127; getting lost in, 132, 134-135; life forms, 126-128, 130-132
Nuthatch, red-breasted, 70
Nutter, John, 76; hike with, 76-77, 80-83, 86-87

O
Oak(s), 26, 53
Old Speck Mountain, 146
Onion, wild, 159
Osprey. *See* Hawk(s)
Otter, 110, 130
Owl(s), 118, 131; barred *(Strix varia)*, *124-125*; great horned *(Bubo virginianus)*, 118, *124*, 160

P
Pamola (storm bird), 148
Parapet Brook, 86
Partridgeberry, 21, 128
Penobscot, West Branch, 148, 153
Peters, Samuel, quoted, 44
Pinchot, Gifford, 55
Pine(s), 53, *175*; pumpkin, 28; white *(Pinus strobus)*, *end paper 2-end paper 3*, 26, 28-31, 157, *162-163*
Pinkham Notch, 76, 83, 87, *96-97*
Pitcher plant: northeastern *(Sarracenia purpurea)*, 60, *61-62*; *Sarracenia flavia*, 62
Pleasant River, West Branch, 147
Porcupine, 21, 110, 112
Porter, Ray, 153
Presidential Range, 45, 72-73, 76-78, 80-84, 88, *102-103*; dangers hiking, 73, 81-82; exploration and development, 83-84, 86; life forms, 77-78, *79*, 80-81, 82, 87, 88; weather, 76, 77, 81-82, 88. *See also* White Mountains and specific peaks

R
Rabbit, 112, 123
Ragweed, 26

Rangeley Lake, 147
Raspberry, 159
Raven, 80
Richardson lakes, 147
Rocks, 46, *47-49;* arkose, *49; Felsenmeer, 102-103;* gneiss, *47,* 50, 51; granite, *4-5,* 46, *49, 84-85, 95, 98-99;* limestone, 46, *47,* 50; marble, 46, *47,* 50, 51; pegmatite, 46, *48;* phyllite, 46, *48;* rhyolite, *175;* sandstone, *49,* 50; schist, *26-27, 49,* 50, 77, 88, 98, *100;* shale, 50, 51, 100; slate, *47, 48*
Rollins, Frank, 84
Ross, John, 148, 149

S
Saddleback Mountain, 146
St. John River, 154, 160, *162-163;* birds, 159-160; name, 154; proposed dam, 165; rapids, 161, 164; trip down, 154, 156-157, 159-161, 164-165
Salamander, red eft *(Diemictylus viridescens), 62*
Sandpiper, 159
Sandwort, Greenland, 77
Sarsaparilla, wild *(Aralia nudicaulis), 68,* 69-70
Sedge(s), 53, 60, *64-65,* 77, 166
Serviceberry, 132
Seven Islands, 161, *162-163*
Sheep, Merino, 53
Shrew, 80, 81; long-tailed, 80; masked *(Snex cinereus), 82;* pygmy, 80; short-tailed, 80; water, 80
Skunk cabbage, *12-13*
Smith, Robert (Smitty), 127; territory of, 127-128, 130-132, 134
Society for the Protection of New Hampshire Forests, 84
Sparrow, white-throated, 82, 87
Spaulding Lake, *94,* 95
Springer, John: *Forest Life and Forest Trees,* 29; quoted, 29, 146
Spruce, 26, *34,* 53, 56, 82, 86, 88, 90, 126, 128, 131, 146, 153, 157, *162-163, 167, 169;* adaptations, 78; black *(Picea mariana), 64, 70, 77, 91,* 131; red, 58, 60, 131; white, 131
Spruce Peak, *56-57*
Squirrel, gray, 21. *See also* Chipmunk
Stiles, Ezra, 26; quoted, 24
Stratton Pond, 56
Sugarloaf Mountain, 146, 147
Sundew *(Drosera rotundifolia), 62*

T
Table Mountain, 88, *98-99*
Taconic Range, 20; formation, 50
Tamarack, 131
Tanner, William, 23-24; cellar hole, *22*
Thoreau, Henry David, 54; quoted, 126, 146, 149
Ticklenaked Pond, 46
Toad, American *(Bufo terrestris americanus), 132-133*
Trees: autumn changes, 164-165; *krummholz,* 78, 90, *91;* of north country, 131-132; persistence of, 24-25; succession, 26, 31, 53; treeline, 77-78, 90; uses, 28-29, 53, 153. *See also* Forests; Logging; and specific types
Trillium, painted, 87
Trout, 161; brook, 128
Tuckerman Ravine, 82, 86
Turkey, wild, 109-110

U
Udall, Stewart, quoted, 55
Underwood, William Lyman, 104-105
Unknown Pond, 130-131

V
Vegetation: above 2,500 feet, 58; adaptation to climate, 78, 80; alpine, 77, 78, *79,* 88, 90, *91;* and animal life, 109; carnivorous, 60, *61, 62,* 64, 65, 67; and ice ages, 53; pollen in cores, 53; subalpine, 78, 88; succession, 25-26, 53
Vermont, 44; deer, 109; economy, 46, 53; Land Use and Development Act 250, 59; name, 50; ski areas, 45-46; topography, 44-45; turkeys, 109. *See also* Green Mountains
Violet, wild white, 69
Vogelmann, Hubert, 58, 60; walk with, 60-70
Vole, 80

W
Walden, 130
Walls, stone, 24
Warbler(s), 87; yellow *(Dendroica petechia),* 118, *120-121;* yellowthroat, 70
Wassataquoik Lake, 152
Wassataquoik Stream, *142-143, 144-145,* 152
Water lily, 60, 64
Water strider *(Gerris marginatus), 64,* 65
Weasel, 81, 110
Weather, 32; on Mount Washington, 72-73, *74-75,* 76, 88; in White Mountains, 88; winter, 32, *33-43*
Weeks, John, 84
Weymouth, George, 28
Whiskey jack. *See* Jay
White Mountain National Forest, 73, 84, 86
White Mountains, *end paper 4-page 1,* 20, 72, 78, 84, 88, *89-103;* climate, 88; exploration and development, 83-84. *See also* Presidential Range
Winooski River, 45
Winthrop, Theodore, quoted, 104
Wolf, 24; Ontario, 112, 115; timber, 108, 109, 127
Wood sorrel, 87
Woodchuck, 80
Woodcock, 109
Woodpecker, 131
Wren, winter, 87

Y
Yellowlegs, 159

Z
Zealand Notch, *92-93*
Zealand Pond, *end paper 4-page 1*
Zealand River, 88, *89, 95*

Printed in U.S.A. XX